T0257161

CLIMATE AND
THE OCEANS

....

Geoffrey K. Vallis

PRINCETON UNIVERSITY PRESS   *Princeton & Oxford*

Published by Princeton University Press
41 William Street, Princeton, New Jersey 08540
In the United Kingdom: Princeton University Press
6 Oxford Street, Woodstock, Oxfordshire OX20 1TW
press.princeton.edu

ISBN: 978-0-691-14467-2 (cloth)
ISBN: 978-0-691-15028-4 (pbk)

Library of Congress Cataloging-in-Publication Data

Vallis, Geoffrey K.
Climate and the oceans / Geoffrey K. Vallis.
p. cm. — (Princeton primers in climate)
Includes bibliographical references and index.
1. Oceanography—Research. 2. Ocean circulation.
3. Ocean-atmosphere interaction. 4. Climatic changes.
I. Title.
GC57.V26 2011
551.5'246—dc22
2011014372

British Library Cataloging-in-Publication Data is available

This book has been composed in Minion Pro

10 9 8 7 6 5 4 3 2 1

Contents

Preface

The truth and nothing but the truth, but not
the whole truth.

THIS IS A BOOK ON CLIMATE, WITH AN EMPHASIS ON THE role of the ocean. The emphasis is on large-scale processes and phenomena, and on the physical aspects of the ocean rather than its chemical or biological properties. It is not a textbook on physical oceanography, of which there are several good ones, nor is it a textbook on climate, of which there are some good ones. Rather, and as its size may indicate, the book is an introduction to, or a primer on, the ocean–climate system.

This book could be used to provide an introductory "big picture" for more advanced students or for scientists in other fields, or it could provide advanced reading for undergraduate students taking courses at a more elementary level. The book is somewhat more mechanistic than most books at this level: the emphasis is on *how things work,* and in particular how the ocean works and how it influences climate. I discuss observations to motivate the discussion, but the main emphasis is not on what things happen to be, but *why* and *how* they happen to be.

This is a *fast* book, although it is not, I hope, a *loose* book. It covers a lot of ground, quickly, and tries not to get bogged down in too much detail. Having said that, one of the most important questions to answer in the study of climate is to understand just what is a detail and what is essential. If one is studying the climate as a whole, then one might regard the presence of a small island in the northeast Atlantic as a detail, and it surely is (unless one is studying the climate of that island). One might also regard the precise way in which carbon dioxide molecules vibrate and rotate when electromagnetic radiation impinges upon them as a detail, yet it is precisely this motion that gives us the greenhouse effect, which makes our planet habitable, and which also gives us global warming, and which may affect our economy to the tune of trillions of dollars. Hardly a detail! But can we, and need we, understand all such matters to understand climate? Without answering that question, it is clear that in a short book such as this, choices have to be made. I will occasionally, but only occasionally, simply tell you how things are, without delving into the mechanisms if they are tangential to our narrative.

Given that we do move around sharp bends quite quickly, a certain amount of sophistication is assumed of the reader, or at least a willingness to think a little and puzzle things out and perhaps even look up one or two references. On the other hand, I haven't assumed much background knowledge—just a bit of basic physics and mathematics and some general knowledge about the ocean and climate. In other words, I'm writing the book for a smart and motivated but somewhat ignorant reader, and I hope you don't mind being so character-ized. In some contrast to most elementary or under-graduate books, I have not shied away from topics that are of current interest or subjects of current research. It should be clear from both the text and the context when the contents of a section are not completely settled or are still a topic of research, rather than being wholly stan-dard textbook material. I have tried to be as objective as possible when discussing such matters, but this attempt does not mean giving two sides of a controversial matter equal weight or suggesting that both are equally valid. Sometimes, an opinion is just plain wrong. The topics of current interest are especially noticeable in the second half of the book, in the chapters and sections on climate variability, global warming, and climate change. The chapter on ocean circulation also reflects our relatively recent understanding of the ocean's meridional over-turning circulation. The more mathematical aspects of the book tend to be concentrated in the first half of the book, and the reader who may be surprised that some of the topics dealing with current research are less math-ematical should ponder the quotation at the opening of chapter 3. Some readers may wish to skip ahead directly to chapters that particularly interest them, and this way of reading should be possible by referring back to the earlier chapters as needed.

I thank Peter Gent and Carl Wunsch for their percep-tive reviews of an early draft of this book. They saved me from myself in a number of ways. Thanks also to

Ryan Rykaczewski, Stephanie Downes, Mehmet Ilicak, Rym Msadek, Caroline Müller, Amanda O'Rourke, Ilissa Ocko, Thomas Spengler, and Antoine Venaille, for many comments and constructive criticisms, even if they did it for the beer. The chapter on ocean circulation owes much to a couple of vigorous conversations with Max Nikurashin. Finally, my thanks to Cathy Raphael for expertly creating many of the figures.

CLIMATE AND THE OCEANS

1 BASICS OF CLIMATE

The climate's delicate, the air most sweet.

—William Shakespeare, *A Winter's Tale*

To APPRECIATE THE ROLE OF THE OCEAN IN CLIMATE, we need to have a basic understanding of how the climate system itself works, and that is the purpose of this chapter. Our emphasis here is the role of the atmosphere—we don't pay too much attention to the oceans as we'll get more of that (lots more) in later chapters—and we assume for now that the climate is unchanging. So without further ado, let's begin.

THE PLANET EARTH

Earth is a planet with a radius of about 6,000 km, moving around the sun once a year in an orbit that is almost circular, although not precisely so. Its farthest distance from the sun, or *aphelion*, is about 152 million km, and its closest distance, *perihelion*, is about 147 million km. This ellipticity, or eccentricity, is small, and for most of the rest of the book we will ignore it. (The eccentricity is not in fact constant and varies on timescales of about 100,000 years because of the influence of other planets on Earth's orbit;

these variations may play a role in the ebb and flow of ice ages, but that is a story for another day.) Earth itself rotates around its own axis about once per day, although Earth's rotation axis is not parallel to the axis of rotation of Earth around the sun. Rather, it is at an angle of about 23°, and this is called the *obliquity* of Earth's axis of rotation. (Rather like the eccentricity, the obliquity also varies on long timescales because of the influence of the other planets, although the timescale for obliquity variations is a relatively short 41,000 years.) Unlike the ellipticity, the obliquity is important for today's climate because it is responsible for the seasons, as we will see later in this chapter.

Earth is a little more than two-thirds covered by ocean and a little less than one-third land, with the oceans on average about 4 km deep. Above Earth's surface lies, of course, the atmosphere. Unlike water, which has an almost constant density, the density of the air diminishes steadily with height so that there is no clearly defined top to the atmosphere. About half the mass of the atmosphere is in its lowest 5 km, and about 95% is in its lowest 20 km. However, relative to the ocean, the mass of the atmosphere is tiny: about one-third of one percent of that of the ocean. It is the weight of the atmosphere that produces the atmospheric pressure at the surface, which is about 1,000 hPa (hectopascals), and so 10^5 Pa (pascals), corresponding to a weight of 10 metric tons per square meter, or about 15 lb per square inch. In contrast, the pressure at the bottom of the ocean is on average about 4×10^7 Pa, corresponding to 4,000 metric tons per square meter or 6,000 lb per square inch!

Table 1.1
Main Constituents of the Atmosphere

Constituent	Molecular weight	Proportion by volume
Nitrogen, N_2	28.01	78.1%
Oxygen, O_2	32.00	20.9%
Argon, Ar	39.95	0.93%
Water vapor, H_2O	18.02	~0.4% (average)
		~1%–4% (at surface)
Carbon dioxide, CO_2	44.01	390 ppm (0.039%)
Neon, Ne	20.18	18.2 ppm
Helium, He	4.00	5.2 ppm
Methane, CH_4	16.04	1.8 ppm

Molecular weight is the molar mass, measured in grams per mole. In addition, there are trace amounts of krypton, hydrogen, nitrous oxide, carbon monoxide, xenon, ozone, chlorofluorocarbons (CFCs), and other gases.

The atmosphere is composed of nitrogen, oxygen, carbon dioxide, water vapor, and a number of other minor constituents, as shown in table 1.1. Most of the constituents are well mixed, meaning that their proportion is virtually constant throughout the atmosphere. The exception is water vapor, as we know from our daily experience: Some days and some regions are much more humid than others, and when the amount of water vapor reaches a critical value, dependent on temperature, the water vapor condenses, clouds form, and rain may fall.

Earth's temperature is, overall, maintained by a balance between incoming radiation from the sun and the radiation emitted by Earth itself, and, slight though it may be compared to the ocean, the atmosphere has a substantial effect on this balance. This effect occurs because water vapor and carbon dioxide (as well as some other minor constituents) are *greenhouse gases*, which means that they absorb the infrared (or longwave) radiation emitted by Earth's surface and act rather like a blanket over the planet, keeping its surface temperature much higher than it would be otherwise and keeping our planet habitable. However, we are getting a little ahead of ourselves—let's slow down and consider in a little more detail Earth's radiation budget.

RADIATIVE BALANCE

Solar radiation received

Above Earth's atmosphere, the amount of radiation, S, passing through a plane normal to the direction of the sun (e.g., the plane in the lower panel of figure 1.1) is about 1,366 W/m^2. (A watt is a joule per second, so this is a *rate* at which energy is arriving.) However, at any given time, half of Earth is pointed away from the sun, so that on average Earth receives much less radiation than this. How much less? Let us first calculate how much radiation Earth receives in total every second. The total amount is S multiplied by the area of a disk that has the same radius as Earth (figure 1.1). If Earth's radius is a, then the area of the disk is $A = \pi a^2$, so that the rate of total radiation received is $S\pi a^2$. Over a 24-hour period, this radiation is spread out over the entire surface of Earth, although not, of course, uniformly. Now, Earth is almost a sphere of

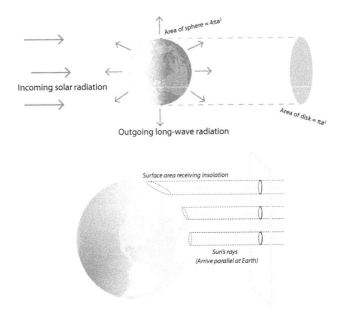

Figure 1.1. Top: The incoming solar radiation impinges on a disk of area πa^2 but is on average spread out over a sphere of area $4\pi a^2$. Bottom: Variation of incoming solar radiation with latitude. A given amount of radiation is spread over a larger area at high latitudes than at low latitudes, so the intensity of the radiation is diminished, and thus high latitudes are colder than low latitudes.

radius a, and the area of a sphere is $4\pi a^2$. Thus, the average amount of radiation that Earth receives per unit area may be calculated as follows:

$$\text{Total radiation received} = S\pi a^2. \tag{1.1}$$

$$\text{Area of Earth} = 4\pi a^2. \tag{1.2}$$

Average radiation received $= \dfrac{S\pi a^2}{4\pi a^2} = \dfrac{S}{4}.$ \hfill (1.3)

That is, the average rate at which radiation is received at the top of Earth's atmosphere is $1{,}366/4 \approx 342$ W/m^2, and we denote this S_0; that is, $S_0 = S/4 \approx 342$ W/m^2.

Distribution of incoming solar radiation

The distribution of solar radiation is obviously not uniform over the globe, as figure 1.1 illustrates. Plainly, low latitudes receive, on average, much more solar radiation than high latitudes, which is the reason why temperatures, on average, decrease with latitude.

The situation is made more complex by the nonzero obliquity of Earth's axis of rotation; that is, the rotation axis of Earth is not perpendicular to its orbital plane, as illustrated in figure 1.2. The axis of Earth's rotation is fixed in space and does not vary as Earth rotates around the sun; that is, relative to the distant galaxies, the line from the South Pole to the North Pole always has the same orientation. But because Earth rotates around the sun, the orientation varies relative to the sun. One day a year, the North Pole is most inclined toward the sun, and this day is known as the Northern Hemisphere's *summer solstice*. These days, it usually occurs on June 20 or 21. The Northern Hemisphere receives much more radiation than the Southern Hemisphere at this time of year, so this corresponds to the Northern Hemisphere's summer and the Southern Hemisphere's winter. In fact, not

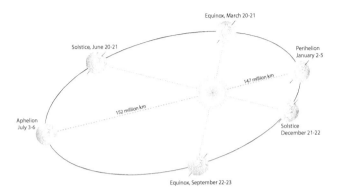

Figure 1.2. Earth's orbit around the sun and the march of the seasons. Earth's axis of rotation is at an angle with respect to the axis of rotation of Earth around the sun. The Northern Hemisphere's summer and the Southern Hemisphere's winter result when the North Pole points toward the sun, and the opposite season occurs six months later. The eccentricity is much exaggerated in the figure.

only is the sun higher in the sky during summer, but the day is also much longer, and at latitudes above the Arctic Circle, the sun does not set for about two weeks on either side of the summer solstice.

Progressing from June on, as Earth moves around the sun, the distribution of solar radiation becomes more equal between the hemispheres, and the length of day evens out. Then we enter autumn in the Northern Hemisphere and spring in the Southern Hemisphere. At the *equinoxes* (about March 20 and September 22), the hemispheres receive equal amounts of radiation from the sun, which is directly above the equator. The Northern Hemisphere's *winter solstice* occurs on December

21 or 22, when the South Pole is most inclined toward the sun. It might seem from this description that in the Northern Hemisphere the coldest day should be on December 21 and the warmest day on or about June 21. In fact, the coldest and warmest times of year occur a few weeks after these dates; the main reason is thermal inertia in the oceans, which delays the onset of the warmest and coldest days. We'll discuss this effect more in chapter 5. Finally, we note that because of the eccentricity of Earth's orbit, Earth's distance from the sun varies throughout the year. However, this variation is a minor factor in seasonality, and for most intents and purposes we can regard Earth's orbit as circular.

A simple radiation model

Let us put aside the spatial variation of solar radiation for a while and try to obtain an estimate of the average surface temperature on Earth, given the average solar radiation coming in at the top of the atmosphere. Solar radiation causes Earth's surface to warm and emit its own radiation back to space, and the balance between incoming and outgoing radiation determines the average temperature of Earth's surface and of the atmosphere. To calculate the temperature, we need to know a few pieces of physics; in particular, we need to know how much radiation a body emits as a function of its temperature and the wavelength of the radiation.

A *blackbody* is a body that absorbs and emits electromagnetic radiation with perfect efficiency. Thus, all the radiation—and therefore all the visible light—that falls upon it is absorbed. So, unless the body is emitting its own visible radiation, the body will appear black. Now, unless it has a temperature of absolute zero (0 K), the blackbody emits radiation and, as we might expect, the amount of this radiation increases with temperature, although not linearly. The amount, in fact, increases at the fourth power of the absolute temperature; that is, the flux of radiation emitted by the body per unit area varies as

$$F = \sigma T^4, \tag{1.4}$$

where $\sigma = 5.67 \times 10^{-8}\,\mathrm{W\,m^{-2}\,K^{-4}}$ is the Stefan–Boltzmann constant. The presence of a fourth power means that the radiation increases very rapidly with temperature. As a concrete illustration, let us suppose that Earth is a blackbody with a temperature of −18°C, or 255 K (which is a temperature representative of places high in the atmosphere). The energy flux per unit area is $F = \sigma \times 255^4 = 240\,\mathrm{W\,m^2}$. The sun, by contrast, has a surface temperature of about 6,000 K, and so the radiation it emits is $F_{\text{sun}} = \sigma \times 6,000^4 = 7.3 \times 10^7\,\mathrm{W\,m^2}$. Thus, although the sun's surface is only about 24 times hotter than Earth, it emits about 300,000 times as much radiation per unit area. It is, of course, the sun's radiation that makes life on Earth possible.

A blackbody emits radiation over a range of wave numbers, but the peak intensity occurs at a wavelength that is inversely proportional to the temperature; this is known as Wien's law. That is,

$$\lambda_{\text{peak}} = \frac{b}{T}, \tag{1.5}$$

where λ_{peak} is the wavelength at peak intensity, b is a constant, and T is the temperature. For T in Kelvin and λ_{peak} in meters, $b = 2.898 \times 10^{-3}$ mK. From equation 1.5, it is evident that not only does the sun emit more radiation than Earth, it also emits it at a shorter wavelength. With $T = 6,000$ K, as for the sun, we find $\lambda_{\text{peak}} = 0.483 \times 10^{-6}$m or about 0.5 μm. Electromagnetic radiation at this wavelength is visible; that is (no surprise), the peak of the sun's radiation is in the form of visible light. (This fact is no surprise because eyes have evolved to become sensitive to the wavelength of the radiation that comes from the sun.) On the other hand, the radiation that Earth emits (at 255 K) occurs at $\lambda_{\text{peak}} = 1.1 \times 10^{-5}$m, which is infrared radiation, also called longwave radiation. The importance of this difference lies in the fact that the molecules in Earth's atmosphere are able to absorb infrared radiation quite efficiently but they are fairly transparent to solar radiation; this difference gives us the greenhouse effect, which we will come to soon.

A simple climate model

We are now in a position to make what is probably the simplest useful climate model of Earth, a *radiation-balance* or *energy-balance model* (EBM), in which the net solar radiation coming in to Earth is balanced by the infrared radiation emitted by Earth. A fraction, α, known as the *albedo*, of the solar radiation is reflected back to space by clouds, ice, and so forth, so that

Net incoming solar radiation = $S_0(1 - \alpha) = 239 \text{ Wm}^2$,   (1.6)

with $\alpha = 0.3$ (we discuss the factors influencing the albedo more below).

This radiation is balanced by the outgoing infrared radiation. Now, of course, Earth is not a blackbody at a uniform temperature, but we can get some idea of what the average temperature on Earth should be by supposing that it is, and so

Outgoing infrared radiation = σT^4. (1.7)

Equating equations 1.6 and 1.7, we have

$$\sigma T^4 = S_0(1 - \alpha),$$ (1.8)

and solving for T, we obtain $T = 255$ K or $-18°$C. For obvious reasons, this temperature is known as the average emitting temperature of Earth, and it would be a decent approximation to the average temperature of Earth's surface if there were no atmosphere. However, it is in fact substantially lower than the average temperature of Earth's surface, which is about 288 K, because of the greenhouse effect of Earth's atmosphere, as we now discuss.

Greenhouse effect

Earth is covered with a blanket of gas made mainly of nitrogen, oxygen, carbon dioxide, and water vapor. This blanket is essential to life on Earth, for (at least) two reasons:

1. We breathe the air, taking in the oxygen and breathing out carbon dioxide. Similarly, plants use sunlight together with the carbon dioxide in the atmosphere to photosynthesize and create organic compounds.
2. The atmosphere absorbs the infrared or longwave radiation emitted by Earth's surface and re-emits it back to Earth, so warming the surface up to a habitable temperature.

Because this is a book on physical science, not biology, we will consider only this second effect, which is similar to the effect of glass in a greenhouse and for that reason is called the *greenhouse effect*.

Let us, then, consider the path of solar and infrared radiation through the atmosphere, as illustrated in figure 1.3. If the sky is clear, then most of the solar radiation incident at the top of the atmosphere goes through the atmosphere to the surface. If the sky is cloudy, then roughly half of the solar radiation is reflected back to space, with the rest either absorbed in the cloud or passing through to Earth's surface. When the solar radiation reaches the surface, a fraction is reflected and returns to space, and the rest of the radiation is absorbed, warming the surface. All told, combining the effects of the reflection by clouds and at Earth's surface, the fraction of solar radiation reflected back to space—that is, the planetary albedo—is about 0.3. The albedo of clouds themselves

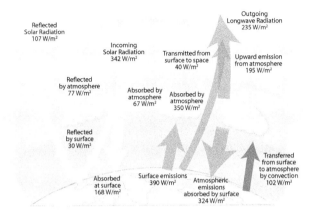

Figure 1.3. The energy budget of Earth's atmosphere, showing the average solar and longwave radiative fluxes per unit area and the convective flux from the surface to the atmosphere. Adapted from Kiehl and Trenberth, 1997.

is higher than this, typically about 0.5, but it can be as high as 0.9 for thick clouds; the albedo of the surface is on average about 0.1 but is much higher if the surface is covered with fresh snow or ice.

The surface is warmed by the solar radiation absorbed and so emits radiation upward, but because the surface temperature of Earth is so much less than that of the sun, the radiation emitted has a much longer wavelength—it is infrared radiation. Now, the atmosphere is *not* transparent to infrared radiation in the same way that it is to solar radiation; rather, it contains greenhouse gases—mainly carbon dioxide and water vapor—that absorb the infrared radiation as it passes through the atmosphere. Naturally enough, this absorption warms the atmosphere, which then re-emits infrared radiation, some of it downward, where it is absorbed at Earth's surface. Thus, and look again at figure 1.3, the total downward radiation at the surface is much larger than it would be if Earth had no atmosphere. Consequently, the surface is much warmer than it would be were there no atmosphere, and this phenomenon is known as the *greenhouse effect*.

A simple mathematical model of the greenhouse effect

Let us now construct a simple mathematical model illustrating the greenhouse effect. Our purpose in doing so is to see somewhat quantitatively, if approximately, whether the atmosphere might warm the surface up to the observed temperature. Let us make the following assumptions:

1. The surface and the atmosphere are each characterized by a single temperature, T_s and T_a, respectively.
2. The atmosphere is completely transparent to solar radiation.
3. Earth's surface is a blackbody.
4. The atmosphere is completely opaque to infrared radiation, and it acts like a blackbody.

The model is illustrated in figure 1.4, which the reader will appreciate is a very idealized version of figure 1.3.

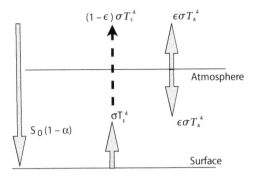

Figure 1.4. An idealized two-level energy-balance model. The surface and the atmosphere are each characterized by a single temperature, T_s and T_a. The atmosphere absorbs most of the infrared radiation emitted by the surface, but it is transparent to solar radiation.

The parameter ϵ, called the emissivity or the absorptivity, determines what fraction of infrared radiation coming from the surface is absorbed by the atmosphere, and we initially assume that $\epsilon = 1$; that is, the atmosphere is a blackbody and absorbs all the surface infrared radiation. The incoming solar radiation, S_0, and the albedo are presumed known, and the unknown temperatures T_s and T_a are obtained by imposing radiative balance at the surface and the atmosphere. At the surface, the incoming solar radiation, $S_0(1 - \alpha)$, plus the downward longwave radiation emitted by the atmosphere is balanced by the longwave radiation emitted by the surface, and therefore

$$S_0(1 - \alpha) + \sigma T_a^4 = \sigma T_s^4. \tag{1.9}$$

Similarly, radiative balance at the top of the atmosphere is

$$S_0(1 - \alpha) = \sigma T_a^4. \tag{1.10}$$

Note that instead of equation 1.10, we could have used the condition that the radiation absorbed by the atmosphere must balance the longwave radiation emitted by the atmosphere so that, looking again at figure 1.4,

$$\sigma T_s^4 = 2\sigma T_a^4. \tag{1.11}$$

Using equation 1.9 and either equation 1.10 or 1.11 and just a little algebra, we obtain expressions for T_a and T_s, namely

$$T_a = \sqrt[4]{\frac{S_0(1 - \alpha)}{\sigma}}, \qquad T_s = \sqrt[4]{\frac{2S_0(1 - \alpha)}{\sigma}}. \tag{1.12a, b}$$

Note that the equation for T_a, namely equation 1.10 is the same as equation 1.8, and so using $S_0 = 342$ Wm^2 and $\alpha = 0.3$, we find $T_a = 255$ K. For the surface temperature, we obtain $T_s = 303$ K $= 30°$C. This temperature is quite a bit higher than the observed average temperature at Earth's surface (288 K) mainly because we are assuming that Earth's atmosphere is a perfect blackbody, absorbing all the longwave radiation incident upon it. In fact, some of the longwave radiation emitted by the surface escapes to space, so let's try to model that in a simple way.

A leaky blanket

Instead of assuming that the atmosphere absorbs all the longwave radiation emitted by the ground, let us assume that it absorbs just a fraction, ϵ, of it, where $0 < \epsilon < 1$; thus, an amount $\sigma T_s^4 (1 - \epsilon)$ of surface radiation escapes to space. Similarly, we also assume that the atmosphere only emits the (same) fraction ϵ of the amount it would do as a blackbody. In other respects, the model is the same as that described above.

The radiative balance equations corresponding to equations 1.9 and 1.10 are, for the surface,

$$\sigma T_s^4 = S(1 - \alpha) + \epsilon \sigma T_a^4 \tag{1.13}$$

and at the top of the atmosphere

$$S_0(1 - \alpha) = \epsilon \sigma T_a^4 + (1 - \epsilon) \sigma T_s^4. \tag{1.14}$$

Solving for T_s, we obtain

$$T_s = \sqrt[4]{\frac{2S_0(1 - \alpha)}{\sigma(2 - \epsilon)}}. \tag{1.15}$$

If $\epsilon = 1$, we recover the blackbody result (equation 1.12b) with $T_s = 303$ K. If $\epsilon = 0$, then the atmosphere has no radiative effect and, as in equation 1.8, we obtain $T_s = 255$ K. The real atmosphere is somewhere between these two extremes, and for $\epsilon = 3/4$, we find $T_s = 287$ K, close to the observed average surface temperature.

Evidently, the surface temperature increases as ϵ increases, and this phenomenon is analogous to what is happening vis à vis global warming. As humankind puts carbon dioxide and other greenhouse gases into the atmosphere, the emissivity of the atmosphere increases. The atmosphere absorbs more and more of the longwave radiation emitted by the surface and re-emits it downward, and consequently the surface temperature begins to rise. The mechanism is relatively easy to understand. It is much harder to determine precisely how much the emissivity rises and how much the surface temperature rises as a function of the amount of carbon dioxide in the atmosphere. We'll discuss this question more in chapter 7, although there is still no exact answer to it. So let's finish off this chapter by discussing a question to which we do have a reasonable answer, namely, which gases in the atmosphere contribute to the greenhouse effect.

What are the important greenhouse gases?

The most important greenhouse gases in Earth's atmosphere are water vapor and carbon dioxide, followed by ozone, nitrous oxide, methane, and various other gases (see table 1.1). Water vapor (and clouds) and carbon dioxide have the dominant effect, although their respective levels are maintained in different ways. The level of CO_2 is almost constant throughout the atmosphere, and year by year its level is maintained by a balance between emissions (e.g., respiration by animals and bacteria) and natural sinks, including photosynthesis in plants and absorption by the ocean. On longer timescales, geological processes such as volcanic outgassing, weathering of rocks, and changes in the ocean circulation come into play, and the CO_2 level may change by natural causes on timescales of thousands of years and longer. As we discuss more in chapter 7, the burning of fossil fuels has led to increased emissions of CO_2 in the past few centuries and the level itself has increased from about 280 ppm in preindustrial times to 390 ppm today.

Water vapor, on the other hand, comes and goes on a daily timescale, as we know from our everyday experience. The source of water vapor in the atmosphere is evaporation from the ocean, lakes, and wet land, and the sink is condensation and rainfall. (Clouds are made mainly of small water droplets that remain suspended in the air.) The level of water vapor varies both spatially (deserts are dry, the tropics are humid) and temporally (some days are dry, others are rainy). But one factor above all determines the average level of water vapor in the atmosphere, and that is temperature. The higher the temperature, the more water vapor a given volume can hold, and so if the atmosphere warms then the amount of water vapor in the atmosphere, on average, increases. Since water vapor is a greenhouse gas, this increase leads to a bit more warming still, and so on, and this process is known as *positive feedback.*

As a rough rule of thumb, in the present climate about half the greenhouse effect comes from water vapor, about a quarter from clouds, and a fifth from carbon dioxide. However, these numbers are approximate because the effects of the greenhouse gases are not additive: if the atmosphere is dry, then adding CO_2 makes a big difference to the greenhouse effect. On the other hand, if we have a lot of water vapor in the atmosphere already so that we already have a significant greenhouse effect, then adding CO_2 does not make nearly as much difference:

Table 1.2

Effect of the main longwave absorbers in the atmosphere.[1]

Absorber	Just the absorber	Everything but the absorber	Range of contribution
Water vapor	62	61	39–62
Clouds	36	85	15–36
Water vapor and clouds	81	33	67–85
Carbon dioxide	25	86	14–25
All others	9	95	5–9
Ozone	5.7	97.3	2.7–5.7
Nitrous oxide	1.6	99	1–1.6
Methane	1.6	99.3	0.7–1.6
Aerosols	1.8	99.7	0.3–1.8
CFCs	0.5	99.9	0.1–0.5

The first two columns of numbers give the approximate percentage of the present greenhouse effect that would remain if either just the absorber or everything but the absorber were present, with temperatures fixed; the third column summarizes the percentage range of the contribution of the absorber. "All others" refers to the combined effects of all other absorbers, which are then listed individually. To obtain radiative fluxes, multiply the percentages by 1.55 W/m^2.

the longwave radiation that CO_2 would absorb has already been absorbed by water vapor. Table 1.2 shows the numbers, which we may interpret as follows. If the only greenhouse gas in the atmosphere were CO_2, then we would have 25% of the greenhouse effect that we have now (that is, a quarter of the long-wave radiation would be absorbed in the atmosphere, if the temperature were held fixed). On the other hand, if we were to remove CO_2 entirely from the atmosphere, the greenhouse effect (longwave absorption) would be reduced by about 14%, to 86% of its current value. If we had only water vapor (and no clouds), we would have about 62% greenhouse effect, and if we remove water vapor but keep everything else, we reduce the present-day greenhouse effect by about 39%. If we had only water vapor and clouds, then we would have 81% of the greenhouse effect, and if we remove water vapor and clouds, then we lose about 67% of the present greenhouse effect.

However, these figures do *not* provide a good measure of the real importance of the dry or noncondensing greenhouse gases (e.g., CO_2, ozone, or methane) because the calculation assumes that the temperature stays the same. If we did not have any of these gases, the temperatures would fall, leading to a reduction in the absolute humidity and a significantly reduced greenhouse effect from water vapor, causing a further reduction of temperature, and so on. In fact, without the dry greenhouse gases, the terrestrial greenhouse effect would likely completely collapse and the temperature would fall so much that Earth would become completely frozen over, forming a "snowball Earth" (Lacis et al., 2010). Thus, ultimately almost all of the greenhouse effect stems from the dry greenhouse gases. The amount of water vapor adjusts to the level of the other greenhouse gases, and so we usually regard water vapor as a feedback and not a primary forcing.

We'll come back to radiative effects in the last chapter, but let's now shift our attention to the ocean and its role in climate, beginning with a descriptive overview of the oceans themselves.

2 THE OCEANS: A DESCRIPTIVE OVERVIEW

Persons attempting to find a motive in this narrative will be prosecuted; persons attempting to find a moral in it will be banished; persons attempting to find a plot in it will be shot.

—Mark Twain, *Adventures of Huckleberry Finn*

WE NOW START TO LOOK AT THE OCEAN(S)[1] IN A LITTLE more detail, albeit in a rather descriptive manner, as a precursor to the more mechanistic or dynamical description, or "explanation," that we try to provide in chapter 4. That is, in this chapter we describe what's going on but with no underlying organizing principle—with no plot, one might say.

There is a sense in which all explanations are really descriptions; what we may think of as an explanation is really a description at a more general level. Nevertheless, the distinction is useful, at least in science: an explanation does not just describe the phenomenon at hand but also provides some more fundamental reason for its properties, and ideally for the properties of a whole class of phenomena. Descriptions are useful because they are the precursor of explanation, and in this chapter our modest goal is to provide a brief descriptive overview of the oceans and their large-scale circulation, focusing on matters that significantly affect climate.

SOME PHYSICAL CHARACTERISTICS OF THE OCEANS

The ocean basins

The ocean covers about 70% of Earth's surface, and so has a total area of about 3.61×10^{14} km^2. Currently, about two-thirds of Earth's land area is in the Northern Hemisphere, so that about 57% of the ocean is in the Southern Hemisphere, 43% in the Northern; the Northern Hemisphere itself is 61% ocean, and the Southern Hemisphere is about 80% ocean. The ocean's average depth is about 3.7 km, but there are deep trenches where the depth reaches about 10 km. The volume of the ocean is approximately 1.3×10^{18} m^3 and, given that the average density of seawater is about 1.03×10^3 kg/m^3, the total mass of the ocean is about 1.4×10^{21} kg, or 1,400,000,000,000,000,000 metric tons. In comparison, the mass of the atmosphere is about 5×10^{18} kg, about 300 times less: the air at the surface weighs about 1,000 times less than seawater, but the effective vertical extent of the atmosphere is a few times greater than that of the ocean.

The ocean basins have changed over time as the continents have moved and deformed as a consequence of the convection in Earth's mantle that leads to the movement of the tectonic plates and so of the continents themselves, all taking place on a timescale of tens to hundreds of millions of years. The ocean itself has existed for a long time, essentially because the water once formed has nowhere to go: the loss of water vapor to space is negligible because nearly all the water vapor is concentrated at the lowest levels of the atmosphere, and the stratosphere and the upper atmosphere are extremely dry. It is believed that the ocean has in fact existed in roughly the sense that we know it now for perhaps some 3.8 Ga (3.8 billion years), since the beginning of the Archaean era, when Earth cooled sufficiently for land masses to form and water to condense. The water itself had its origins both in volcanism and degassing from Earth's interior, and in collisions with extraterrestrial bodies—probably mainly small icy protoplanets (moonlike bodies) and comets. Such collisions were fortunately much more common in this stage in the evolution of the solar system than they are now. Since that time, the ocean basins have certainly come and gone many times. As continents move significantly on a timescale of tens to hundreds of millions of years, one can envision several distinct continent–ocean configurations over Earth's history, and some of the more recent ones are illustrated in figure 2.1 since the breakup of the "supercontinent" Pangea some 200 million years ago. Reconstruction of the configuration naturally becomes increasingly difficult and so more prone to error the further back one goes in time, but it is believed that there may have been a number of supercontinents over Earth's history, perhaps each a few hundred million years apart.

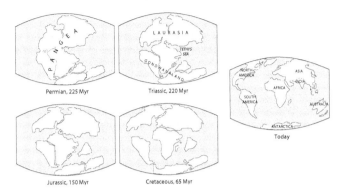

Figure 2.1. Schematic of the configuration of the oceans and continents over the past 225 million years, since the breakup of the supercontinent Pangea. Source: Adapted from USGS (http://pubs .usgs.gov/gip/dynamic/historical.html).

Composition

In today's climate the oceans are mainly liquid; only about 2% of the water on the planet is frozen. Most of the frozen water is in the ice sheets of Antarctica (with 89% of the world's ice, and an average depth of about 2 km) and Greenland (8%, 1.5 km deep). The volume of sea ice, formed by the freezing of seawater, is far less than that of land ice because typically it is only a few meters thick. Its extent also varies considerably by season. However, the importance of land ice and sea ice for climate is in some ways comparable because their areal coverage is similar: about 10% of land is covered with ice year round and about 7% of the ocean. Ice on both land and the ocean has a high albedo, up to 70% when fresh compared to about 10% for seawater, and both make a noticeable difference to the climate at high latitudes.

Seawater itself is made of water plus a collection of minerals, or "salts," mainly chloride (19⁰/₀₀, or 19 parts per thousand by weight) and sodium (11⁰/₀₀). The total average concentration of such salts in the ocean is about 34.5⁰/₀₀ or 34.5 g/kg.[2] Thus, when 1 metric ton (about 1 m³) of seawater evaporates, it yields almost 35 kg of salt (and 1 ft³ yields 2.2 lb of salt). The origin of the salt is weathering and erosion of rocks, as well as the outgassing of chloride from Earth's interior and, when the oceans formed, the leaching of sodium from the ocean floor. The overall level of salinity in the open ocean is now almost constant in time—at least on millennial and shorter timescales—but varies spatially, from a high of about 37⁰/₀₀ in surface waters of the subtropics, where evaporation removes freshwater, to a low of about 32⁰/₀₀ at high northern latitudes, where rain brings freshwater and evaporation is small.

Variations in salinity and temperature bring about corresponding variations in the density of seawater, and these variations play a large part in ocean circulation. Density decreases as temperature increases and increases as salinity increases (although freshwater, but not salty seawater, has anomalous behavior in that its density increases as the temperature rises between 0°C and 4°C). For small variations of temperature, salinity, and density (T, S, and ρ, respectively) around a reference state (T_0, S_0, and ρ_0), density varies according to the formula

$$\rho = \rho_0 [1 - \beta_T (T - T_0) + \beta_S (S - S_0)], \tag{2.1}$$

....

with $\rho_0 = 1.027 \times 10^3$ kg m^{-3}, $T_0 = 10°C$, $S_0 = 35$ g kg^{-1}. The parameters β_T and β_S are the coefficient of thermal expansion and the coefficient of haline contraction, respectively. Their values are not in fact constant throughout the ocean but vary with temperature and pressure, with β_T varying from about 1×10^{-4} K^{-1} (at very low temperatures) to about 2.5×10^{-4} K^{-1} (at high temperatures). The value of β_S varies less and typically is about 8×10^{-4} kg/g. The fact that density varies less with temperature when the water is very cold means that salinity plays a greater role in determining density at low temperatures—and so at high latitudes and in the cold climates of the past—than does temperature itself.

The specific heat of seawater is about 4,180 Jkg^{-1} K^{-1}, which is about four times that of air per unit mass, so overall the heat capacity of the ocean is more than 1,000 times that of the atmosphere. In fact, the top 3 m of an ocean column have about the same heat capacity as the corresponding atmosphere above. This heat capacity is responsible for one of the largest effects of the ocean on climate—its moderating effect in bringing milder winters and cooler summers to locations near the ocean, especially on the western edges of the continental land masses, and in slowing down the progression of global warming, as we discuss in later chapters.

OCEAN STRUCTURE AND CIRCULATION

The aspect of the ocean that perhaps most affects the climate is the temperature distribution at the surface—the

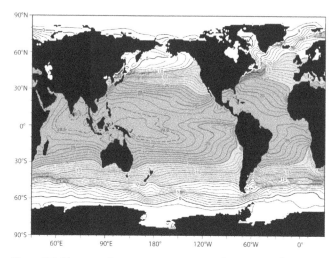

Figure 2.2. The annual average temperature at the ocean surface, in degrees centigrade. Adapted from World Ocean Atlas, 2009 of the National Oceanic and Atmospheric Administration (http://www.nodc .noaa.gov/OC5/WOA09/pr_woa09.html).

sea-surface temperature or SST—which is shown in figure 2.2. We see, as expected, a decrease of temperatures poleward but also a noticeable east–west variation in the midlatitudes. Let us first describe the circulation giving rise to this in a rather basic way, and to that end it is useful, if perhaps a slight oversimplification, to divide the ocean circulation into two components:

1. a quasi-horizontal circulation that is largely driven by the effects of the wind and that gives rise most noticeably to the great ocean *gyres* and

2. a meridional overturning circulation (MOC), in which cold, dense water sinks at high latitudes before moving equatorward and rising in low latitudes and/ or in the opposite hemisphere. (Meridional generically means in the north–south direction.)

How the winds and density gradients produce these currents is the subject of chapter 4; here we are less ambitious and seek simply to describe the circulation and the horizontal and vertical structure of the ocean.

The horizontal structure and the ocean currents

A schematic of the main horizontal ocean currents at the ocean surface is shown in figure 2.3. In most regions of the world, these currents extend a few hundred meters into the ocean, with the exception of the equatorial region. Here there is a shallow westward-flowing surface current and a more substantial eastward flow beneath (as well as narrow countercurrents on either side of the westward flow, not shown in the schematic). At first glance, the circulation seems complicated, especially as the currents interact with the geography of the continental land masses. However, if we look more closely, the circulation actually simplifies, and we can identify a number of features common to each of the major ocean basins.

In mid- and high latitudes, the most conspicuous aspects of large-scale circulation are the circulating *gyres*. Between latitudes of about 15° and 45°, in either hemisphere,

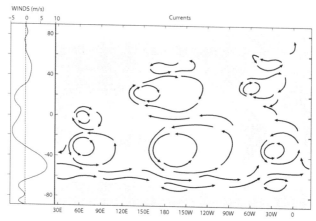

Figure 2.3. A schematic of the main surface currents of the world's oceans. The panel at the left shows the zonally averaged zonal (i.e., east–west) surface winds.

the circulation is dominated by the *subtropical gyres*. The flow in the subtropical gyres is westward on their equatorial branch and eastward in midlatitudes on their poleward branch, and it is not too difficult to imagine that this circulation is directly driven by the wind—the westerlies (i.e., wind from the west) in midlatitudes and the generally easterly (wind from the east) trade winds in low latitudes. There are subtropical gyres in both the Atlantic and Pacific oceans in both hemispheres (figure 2.3), as well as in the southern Indian Ocean. The circulation in these gyres occurs mainly in the upper few hundred meters of the ocean, weakening and eventually becoming unrecognizable in the deep abyss.

A rather conspicuous aspect of the subtropical gyres is their east–west asymmetry: the meridional component of all of these gyres is much stronger in the west of the oceans, giving rise to what are called *western boundary currents*. In the North Atlantic, this current is called the Gulf Stream, in the North Pacific, the Kuroshio, in the South Atlantic, the Brazil Current, in the South Pacific, the East Australia Current, and in the Indian Ocean, the Agulhas Current. The cores of these currents are often only 100 km or so wide, and the speed of the flow can reach a quite tidy 1 m/s, which is not at all sluggish for the ocean, where most currents are more like a few centimeters per second. The equatorward return flow in all the subtropical gyres is spread over a much greater longitudinal extent and so is much weaker.

In the Northern Hemisphere, poleward of the subtropical gyres, the circulation consists of *subpolar gyres*. Because of the converging meridians and the complicated geography of the North Atlantic and North Pacific, these gyres are not nearly as conspicuous as their subtropical counterparts, but they too are primarily wind driven, a consequence of the strong midlatitude westerly winds and the weak easterly winds at high latitudes. They also have intense western boundary currents, now flowing equatorward, and it may be useful to look ahead to see a schematic of how the ocean gyres might be if the oceans were purely rectangular in figure 4.1 in chapter 4. The reader may imagine how the circulation would be distorted, but might keep its essential structure, if the rectangle were replaced by the more complicated geometry of the real ocean basins.

The Antarctic Circumpolar Current

At high latitudes, the Southern Hemisphere has a qualitatively different geometry than the Northern Hemisphere: the other continents do not connect to Antarctica, and so the oceans are free to circulate all the way around the globe, forming the *Antarctic Circumpolar Current* (ACC). The flow is predominantly zonal (east–west) and in that sense resembles that of the atmosphere. An enormous transport of water is sustained over great lengths, comparable to that of the great gyres. The strength of ocean currents is traditionally measured in sverdrups (Sv), which are named after Harald Sverdrup, a famous Norwegian oceanographer working in the first half of the twentieth century. A sverdrup was originally defined as a volume transport of $10^6 \, \mathrm{m}^3 \, \mathrm{s}^{-1}$ of water, although nowadays it is also commonly used as a measure of mass flow, namely $10^9 \, \mathrm{kg} \, \mathrm{s}^{-1}$ of water or 1 million metric tons of fluid per second. (The two definitions are not exactly equivalent because the density of water is not exactly $1{,}000 \, \mathrm{kg} \, \mathrm{m}^{-3}$, but in any case, the sverdrup is usually used in an informal, approximate way.)

The ACC has an average flow rate of about 120 Sv and goes up to 150 Sv in places. In comparison, the Gulf Stream varies from about 30 Sv off the coast of Florida, increasing in strength as it flows northward and water joins it from the Atlantic, eventually peaking at about 150 Sv after it leaves the coast near Cape Hatteras in North Carolina. To give some perspective, the flow of the mighty Amazon River is about 0.2 Sv, and the flow of all the world's rivers into the ocean totals about 1 Sv, whereas the westerly winds in the atmosphere carry up to 500 Sv of air.

The overturning circulation and the vertical structure

The *meridional overturning circulation* (MOC) of the ocean is the name commonly given to the circulation in the meridional (i.e., north–south and up–down in water depth) plane. Horizontal variations in this circulation can be important, but let us put them aside for now. This circulation comes about as a consequence of various factors: the temperature gradient between equator and pole, the temperature difference between high northern latitudes and high southern latitudes, the winds, especially over the southern ocean, and turbulent mixing in the ocean interiors; we will discuss all of these more in chapter 4. Suffice it to say now that the dense water at high latitudes sinks and moves equatorward in the deep ocean, filling the abyss in both hemispheres with water that is very cold and dense. The overturning circulation is much stronger in the Atlantic than in the Pacific because the water at high northern latitudes is saltier, and hence denser, in the Atlantic than in the Pacific, but in both basins the deep water is cold and dense. However, the upper ocean at lower latitudes contains warmer water circulating in the gyres, and this water extends down almost a kilometer deep. The upshot is an

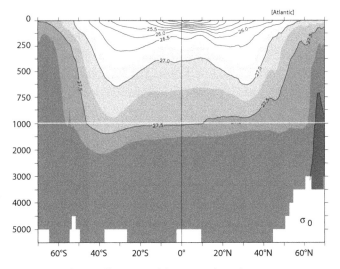

Figure 2.4. The zonally averaged density in the Atlantic Ocean. Note the break in the vertical scale at 1,000 m.[3]

ocean structure as illustrated in figure 2.4, with warm, light water in the subtropical gyres literally floating on top of the cold, dense abyssal water that has come from high latitudes. A typical vertical profile passing through the subtropical gyre might look something like the schematic shown in figure 2.5. We may identify three distinct features: the mixed layer, the thermocline, and the abyss, so let's talk about these features now.

The mixed layer

The mixed layer is the topmost layer of the ocean, in which the water is (no surprise) well mixed, so that its

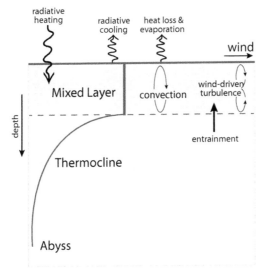

Figure 2.5. Schematic of the vertical structure of the ocean, emphasizing the mixed layer. In the mixed layer, typically 50–100 m deep, turbulence and convection act to keep the temperature relatively uniform in the vertical. Below this layer, temperature changes over a depth of a few hundred meters, in the *thermocline*, before becoming almost uniform at depth, in the *abyss*. Adapted from Marshall and Plumb, 2007.

temperature and salinity are almost uniform with depth. The mixing comes about through mechanical stirring from the wind and by convection that occurs when the water becomes statically unstable—that is, when a denser patch of water lies on top of a lighter patch of water. The differences in density normally arise because of the temperature differences (warmer water is light) but might also arise because of salinity effects (saltier water is denser), for example, if evaporation produces salty water at the surface. Typically, the mixed layer is about 50–100 m thick, but in places it might go much deeper if there is intense convection, which mostly occurs in high latitudes.

The thermocline and the abyss

The *thermocline* is the layer of water in which temperature varies quite rapidly from the warmth of the mixed layer to the cold of the abyss, and typically it is 500–1,000 m thick. The *abyss* is a very thick layer of water that stretches from the base of the thermocline to the bottom of the ocean. The water in the abyss has its origins at high latitudes in both Northern and Southern hemispheres and so is cold and dense. The thermocline may thus be regarded as a transition region, or boundary layer, connecting the cold abyss with the warm surface layers. It is a rather complicated transition region because a number of interacting physical processes occur there. First, the thermocline has a seasonal component because the temperature in the mixed layer varies with the seasons, whereas the abyssal temperature is almost constant year round. The temperature difference across the thermocline is thus larger in summer than in winter. In the upper thermocline, the seasonal variations manifest themselves most. This region is called the *seasonal thermocline,* and it may be several tens of meters thick. The lower, unchanging part of the thermocline is called the *permanent thermocline,* and this part is up to several hundred meters thick. The thermocline is not just a static transition between cold and warm waters; rather, it contains the circulating waters of the great gyres. Finally, we note that the thermocline is rather different in the subpolar gyres than in the subtropical gyres. Typically, it is weaker in the former because the surface waters are already quite cold there, so there is less of a transition region—that is, a much less distinct thermocline. Indeed in some places in the subpolar gyres, the mixed layer is very deep, with well-mixed convective regions extending well into the abyss. Nevertheless, the upper waters are still circulating and over most of the subpolar gyre there is still a thermocline, albeit a somewhat weak one.

GLOBAL OCEAN CIRCULATION AND OCEAN EDDIES

Let us conclude this chapter with a short discussion of two important and rather different aspects of the ocean: if and how the ocean circulates as a whole and, at the opposite extreme, the role of *mesoscale eddies.*

The global circulation and the "conveyor belt"

The ocean contains two major interhemispheric basins, containing the Pacific and the Atlantic oceans, as well as the smaller, primarily Southern Hemisphere, basin containing the Indian Ocean, all connected via the Antarctic Circumpolar Current (ACC). Wind-driven gyres

exist in all the major basins, and the Atlantic has a robust overturning circulation. How does it all fit together?
There is no universally accepted picture, and certainly no
quantitative theory. Rather, observations and numerical
simulations have been used to put together a qualitative
picture, as illustrated in figure 2.6. The circulation illustrated should be regarded as a highly schematic representation, perhaps even as a metaphor, of the real ocean
circulation, in part because of the presence of ocean eddies discussed below. Also, only the more global features
are illustrated; thus, we see a cross-hemispheric circulation in the Atlantic, but not its vertical structure, which
we'll talk about more in chapter 4. However, the figure
does illustrate the main features of the global circulation:
a meridional overturning circulation, the sinking of cold
dense water in the North Atlantic and off Antarctica, and
the western boundary currents.

Ocean eddies

We are all familiar with the fact that the weather differs
from the climate, the difference arising because the atmospheric flow is unsteady, and we talk more about this
in chapter 6. The same applies to the ocean, only more
so: the large-scale currents in the ocean are almost all
unstable, rather like a river flowing over rapids, and tend
to break up into smaller *mesoscale eddies*, as illustrated
in the lower panel of figure 2.6. The resulting eddies are
the oceanic analogue of atmospheric weather, although
because of differences in the physical properties of the

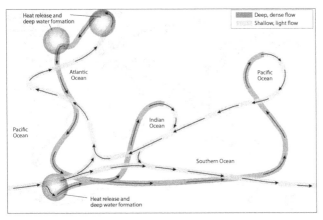

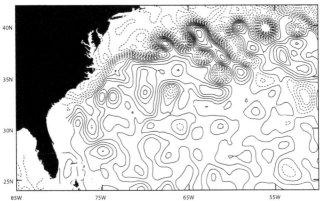

Figure 2.6. Top: An artist's impression of the global ocean circulation, sometimes called the "conveyor belt." Bottom: The sea-surface height in the Atlantic on October 15, 2008, indicating the presence of the Gulf Stream and mesoscale eddies.

two systems ocean eddies tend to be much smaller, with scales of 50–300 km, compared to 500–3,000 km in the atmosphere. In spite of their small scale, the ocean eddies are energetic—the total kinetic energy associated with the eddies in the ocean is about ten times larger than the kinetic energy in the mean currents. Nevertheless, the eddies do not completely destroy the mean flows or make them meaningless, although they can obscure them from easy view and straightforward measurement. In fact, even though eddies dominate the energy budget, they don't dominate the global-scale transport of important properties (such as heat and salinity), and recognizable large-scale oceanic flows remain as a residual after appropriate averaging. Understanding how the eddies and the large-scale mean currents interact remains a daunting challenge in physical oceanography, and the reader may wish to contemplate for a moment the two views of the ocean, juxtaposed in figure 2.6. A full reconciliation of these contrasting views and an understanding of how they fit together are perhaps now coming into our reach, although tantalizingly beyond our grasp.

3 A BRIEF INTRODUCTION TO DYNAMICS

···

Mathematics is the easiest bit in physics.

—Pierre-Gilles de Gennes, *Les Objets Fragile*

WE NOW BEGIN OUR QUEST OF PROVIDING AN EXPLANA-
tion for how and why the ocean circulates the way it does
and how and why it affects the climate. In this chapter,
we'll explain some of the basic dynamical principles
that determine the circulation; in the next chapter, we'll
apply these principles to the circulation itself. The current chapter is a little more mathematical than the others
in this book, but it requires no more sophistication on
the part of the reader—perhaps less in fact. Nevertheless,
some readers may prefer to read the later chapters first,
referring back to the material here as needed.

The atmosphere and ocean are both fluids and, although one is a gas and the other a liquid, their motion is
determined by similar physical principles and described
by similar equations: the *Navier–Stokes equations* of fluid
dynamics. These are complex, nonlinear, partial differential equations that require the largest supercomputers
to solve, but embedded within them are two important
principles that reflect the dominant force balances in the

atmosphere and ocean, namely *hydrostatic balance* and *geostrophic balance.* Hydrostatic balance represents the balance in the vertical direction between the pressure force and the gravitational force, and geostrophic balance represents the horizontal balance between the pressure force and the so-called Coriolis force, a force that arises because of the rotation of Earth. If we understand these forces, we will be able to understand a great deal about the motion of the atmosphere and ocean, so let's figure out what they are. We'll begin with the forces that arise (or appear to arise) as a consequence of Earth's rotation, namely centrifugal force and the Coriolis force, and then consider the pressure force. It turns out that for most geophysical applications, the Coriolis force is much more important than centrifugal force, but we need to understand the latter to understand the former, so that is where we begin.

CENTRIFUGAL FORCE

Suppose that you are riding in a train that starts to go around a bend rather quickly. You feel like you are being thrust outward toward the side of the car, and if you are really going quickly around a tight curve, you might have to hang onto something to stay put. The outward force that you are feeling is commonly known as *centrifugal force.* Strictly speaking, it is not a force at all (we'll explain that cryptic comment later), but it certainly feels like one. What is going on?

One of the most fundamental laws of physics, Newton's first law, says that, unless acted upon by a force, a body will remain at rest or continue moving in a straight line at a constant speed. That is, to change either direction or speed, a body must be acted upon by a force. Thus, in order for you to go around a bend, a force must act (and act on the train too), and this force, whatever it may be in a particular situation, is called the *centripetal force*. Without that force, you would continue to go in a straight line. The centrifugal force that you feel is caused by your inertia giving you a tendency to try to go in a straight line when your environment is undergoing a circular motion, so you feel that you are being pushed outward. You do end up going around the bend because your seat pushes against you, providing a real force (the aforementioned centripetal force) that accelerates you around the bend. The centripetal force that makes the train go around the bend comes from the rails pushing on the train wheels. With this discussion in mind, we see that there are two ways to think about the force balance as you go around the bend (literally).

1. From the point of view of someone standing by the side of the tracks, you are changing direction and a real force is causing you and the train to change direction, in accord with Newton's laws.
2. From your own point of view, you are stationary relative to the train. If you don't look out of the window, you don't know you are going around a bend. There appear to be two forces acting on you: the centrifugal force pushing you out and the seat pushing back (the centripetal force) in the other direction. In this

frame of reference, Newton's law is satisfied because the two forces cancel each other out. That is, in the train's frame of reference you remain stationary and there is a balance of forces between the centripetal force from the seat pushing you in and the centrifugal force pushing you out.

The centrifugal force is, therefore, really a device that enables us to use Newton's laws in a rotating frame of reference: we can say that Newton's laws are satisfied in a rotating frame provided we introduce an additional force, the centrifugal force. There is a simple formula for this force that is derived in appendix A to this chapter. If an object of mass m is going around in a circle of radius r with a speed v, then the centrifugal force, F_{cen}, is given by

$$F_{cen} = \frac{mv^2}{r}. \tag{3.1}$$

The centrifugal force per unit mass is just v^2/r.

A quite analogous situation occurs for us on Earth. Earth is actually rotating quite quickly; it goes around once a day, and the velocity of Earth's surface at the equator is a quite respectable 460 m s^{-1}, or a little faster than 1,000 miles per hour. Sitting on the surface of Earth, we must therefore experience a centrifugal force that is trying to fling us off into space. The reason that we stay comfortably on the surface is that the force of gravity overwhelms the centrifugal force, as a quick calculation shows. The radius of Earth is about 6,300 km, so that using equation 3.1, the centrifugal force per unit mass is given by

$$\frac{460^2}{6.3 \times 10^6} = 0.03 \text{ m s}^{-2}. \tag{3.2}$$

This value should be compared to the force of gravity per unit mass (i.e., the gravitational acceleration) at Earth's surface, which is 9.8 m s^{-2}. The centrifugal effect is therefore quite small, although not so small that we cannot measure it. In fact, because the centrifugal effect is largest at the equator and diminishes to zero at the poles, over time Earth has developed a slight bulge at the equator such that the line of apparent gravity (the real gravity plus the centrifugal force) is perpendicular to Earth's surface. (The distance from the center of Earth to the surface is in fact some 30 km larger at the equator than at the poles.) The centrifugal force is otherwise not terribly important, and its effect can be taken into account by slightly modifying the value of the gravitational force as necessary.

THE CORIOLIS FORCE

Another apparent force is caused by Earth's rotation, one that only arises when bodies are in motion relative to the rotating Earth, and this force is known as the *Coriolis force* after the French engineer and scientist Gaspard-Gustave Coriolis (1792–1843). It turns out to be much more important than the centrifugal force for currents and winds, although its effects are rather subtle. The sphericity of Earth is not important in the Coriolis force itself, and until we get to the section on differential rotation and Earth's sphericity in this chapter, we can imagine Earth to be a rotating disk, with the rotation axis through the North Pole at the center of the disk.

To gain an intuitive idea of what the coriolis force is, consider the (hopefully fanciful) situation in which a missile is launched from the North Pole toward the equator, as illustrated in figure 3.1.[1] Once launched and above Earth's atmosphere, the missile is uninfluenced by the fact that Earth is rotating beneath it. Suppose the missile is initially aimed at Africa and that it takes about six hours for the missile to reach the equator. When the missile reaches the equator, Earth has rotated a quarter turn, but the missile has not and so it will land in South America! From the perspective of someone tracking the missile from the surface of Earth, the missile has not gone in a straight line but has veered to the right.

If a missile is fired from the equator toward the North Pole, it begins its flight with a large eastward velocity, equal to that of the surface of Earth at the equator. As the missile moves poleward, it maintains an eastward velocity (in fact, it conserves its angular momentum), which soon exceeds that of Earth beneath it. From the point of view of an observer on Earth's surface, the missile again appears to veer to the right. From the point of view of an observer on Earth, it seems that the missile has experienced a force—the Coriolis force—that causes it to veer to the right. Just as with the centrifugal force, the Coriolis force is something that we introduce to be able to use Newton's laws in a rotating frame of reference. It is not a real force in the sense that no other body causes it;

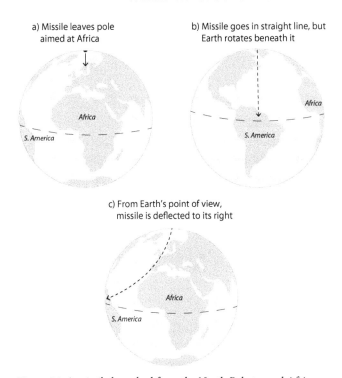

a) Missile leaves pole aimed at Africa

b) Missile goes in straight line, but Earth rotates beneath it

c) From Earth's point of view, missile is deflected to its right

Figure 3.1. A missile launched from the North Pole toward Africa. Earth rotates beneath the missile, and the missile lands in South America. From the point of view of an observer on Earth, the missile has been deflected to its right, and the force causing that deflection is the Coriolis force.

rather, it is a manifestation of the inertial tendency of a body to go in a straight line while Earth rotates.

Let us now imagine that a missile is fired along a line of latitude, neither toward the equator nor away from it. First consider the situation in which the missile is fired in the direction of Earth's rotation. The missile is now rotating around Earth's axis of rotation faster than Earth beneath it, and so the outward centrifugal force on the missile exceeds that of an object stationary relative to Earth. Thus, the missile veers outward from the axis of rotation and so to the right of its original path. Similarly, if the missile is fired in a direction opposite to Earth's rotation, it rotates more slowly than a stationary object and so experiences a weaker centrifugal force than when it was sitting stationary on the ground. It is thus drawn inward toward the axis of rotation, and again appears to be deflected to the right relative to its direction of travel. Indeed, no matter what the missile's initial orientation, when it is in the Northern Hemisphere, it will always veer to the right; similarly, in the Southern Hemisphere, the deflection and the apparent force is always to the left of the direction of motion. Thus, in both hemispheres, a body moving away from the equator veers eastward and a body moving eastward veers toward the equator.

Magnitude of the Coriolis force

There is a simple way we can calculate the Coriolis force on a body moving zonally (i.e., in the east–west direction) if we already know the form of the centrifugal force. For simplicity, we first suppose that Earth is a flat disk, with the axis of rotation perpendicular to the disk and passing through the disk's center, which is then analogous to the North Pole. Gravity points down into the disk. (We consider the effects of sphericity in the next section; until then, Earth is flat.) Let Earth's angular velocity be Ω. (Earth rotates around its axis about once a day, so its angular velocity is 2π radians per day and thus $\Omega \approx 7.27 \times 10^{-5}$ radians per second.) The velocity of the surface of Earth is Ωr, where r is the distance from the axis of rotation, so that a missile moving along a line of latitude (i.e., around the disk) with velocity u relative to Earth has a total velocity, U, of $\Omega r + u$. The total centrifugal force per unit mass experienced by the missile is then given by

$$\frac{U^2}{r} = \frac{(\Omega r + u)^2}{r} = \Omega^2 r + \frac{u^2}{r} + 2\Omega u. \tag{3.3}$$

The first term on the right-hand side, $\Omega^2 r$, is the centrifugal force due to the rotation of Earth itself. The second term is the additional centrifugal force due to the additional velocity of the missile. The third term, $2\Omega u$, is the Coriolis force; for oceanography and meteorology, it is the most important term of the three. Why so? The first term, $\Omega^2 r$ is a constant, and as we noted previously, its effects on Earth can be incorporated into a slightly changed gravitational term. The ratio of size of the second term to the third term is $u{:}2\Omega r$, and given that $2\Omega r \approx 900$m s^{-1}, the ratio is almost always much smaller than unity for winds and ocean currents. The factor 2Ω arises so frequently that we give it the special symbol f, so that the Coriolis force, per unit mass, on a body is equal to f times its speed. The Coriolis force on a body initially moving meridionally (toward the axis of rotation) is given by the same expression (we show this expression explicitly in appendix A to this chapter), with the force acting to deflect the flow so that it begins to move in a zonal direction—in other words, *the Coriolis force acts at right angles to the direction the body is moving.*

It is traditional in oceanography and meteorology to denote the zonal (eastward) velocity of the fluid by the symbol u, and the meridional (northward) velocity by v. Thus, the Coriolis force is given by

Zonal Coriolis force $= fv,$ (3.4a)
Meridional Coriolis force $= -fu.$ (3.4b)

Why is there a minus sign in the second equation? It is because a flow with a positive zonal velocity is deflected to the right in the Northern Hemisphere, toward the equator. The Coriolis force must therefore be negative to generate a negative meridional velocity.

Differential rotation and Earth's sphericity

Finally, let us consider the effect of Earth's sphericity on the Coriolis force. The rotation axis of Earth is a line between the North Pole and the South Pole, and the Coriolis force always acts in a direction perpendicular to this line. We are primarily interested in the *horizontal* deflection of fluid parcels by the Coriolis force because in the vertical direction, gravitational effects dominate and the Coriolis force is relatively small. The horizontal deflection is caused mainly by the component of the rotation that points in the *local vertical* direction acting on the horizontal fluid flow. (There is also a horizontal force caused by the horizontal component of the rotation acting on the vertical fluid motion, but because vertical velocities are small, this force is small also.) If Earth is rotating with an angular velocity Ω, then at a latitude ϑ the component of rotation in the local vertical is equal to $\Omega \sin \vartheta$ (figure 3.2). This angle is equal to Ω at the North Pole but diminishes as one moves equatorward, falling to zero at the equator itself. In the Southern Hemisphere, the component in the local vertical direction decreases from zero at the equator to $-\Omega$ at the South Pole. So here the magnitude of the Coriolis force still increases with latitude, but the force acts in the opposite direction and tends to deflect objects to the left.

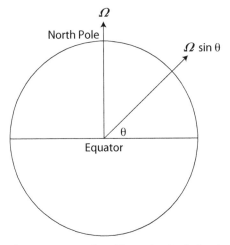

Figure 3.2. The component of Earth's rotation in the local vertical direction varies with latitude (θ) like $\Omega \sin \theta$. Its value is $+\Omega$ at the North Pole, zero at the equator, and $-\Omega$ at the South Pole. The Coriolis parameter f is given by $f = 2\Omega \sin \theta$.

At the poles, Earth's surface is perpendicular to the axis of rotation; horizontal motion is therefore also perpendicular to the axis of rotation and so experiences the full effects of Earth's rotation, just like flow on a rotating disk. At the equator, a horizontal motion does not involve any motion toward or away from the axis of rotation, and the Coriolis force must involve vertical motions, which are small for large-scale flow. That is, at the equator the only components of the Coriolis force that are nonzero either involve the vertical velocity (which is small) or act in the vertical direction (and in this direction, the gravitational force swamps the Coriolis force).

The "effective" rotation of Earth thus increases as we move poleward, and thus Earth's atmosphere and ocean are said to be in *differential rotation*. We take this effect into account by allowing the Coriolis parameter, f, to equal twice the vertical component of the rotation. Thus, $f \equiv 2\Omega \sin \vartheta$, and it increases from the South Pole (where $f = -2\Omega$) to the equator (where $f = 0$) and to the North Pole (where $f = 2\Omega$). Equations involving the Coriolis force, such as equation 3.4a, all hold with this new definition of f. The variation of the Coriolis parameter turns out to be crucial in the production of western boundary currents like the Gulf Stream, as we see in the next chapter.

THE PRESSURE FORCE

A fluid, either a gas or a liquid, is composed of molecules in motion. The collective motion of the molecules gives rise to the flow of the fluid—the winds of the atmosphere and the currents of the ocean—even when there is no such organized flow, the molecules are moving, but their motion is more random. In a gas at room temperature, the molecules are moving at typical speeds of 450–500 m s^{-1}. These molecules naturally collide with each other (in fact, a typical distance a molecule travels before colliding with another is only about 7×10^{-8} m, and this distance is covered in about 0.14 nanoseconds) and with the walls of any enclosing container, and these collisions are the origin of the pressure force in a gas. In a liquid, the molecular motion is not nearly so random, but nevertheless there is a similar pressure force. The pressure force acts not only on the walls of the container but also on the fluid itself, and this force may cause the fluid to move.

If the pressure in a fluid is uniform everywhere, then there is no *net* force. To see this, imagine a small, neutrally buoyant slab floating within the fluid, as illustrated in figure 3.3. There is a pressure force on the left side pushing the slab to the right and a pressure force on the right side pushing it to the left. If the pressure is the same everywhere (that is, $\delta p = 0$ in the figure), these two forces cancel out and the body remains stationary, but if there is a pressure *gradient* ($\delta p \neq 0$), then there is a net force on the body and it will move. If we imagine removing the body and replacing it with the fluid itself (so that we may think of the new piece of fluid as floating in the rest of the fluid) then, again, if there is a pressure gradient, there is a net force on the fluid and, in the absence of opposing forces, the fluid accelerates.

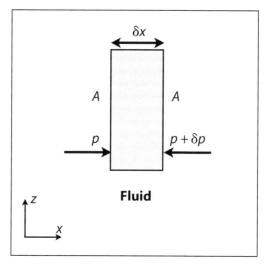

Figure 3.3. A slab (dark shading) floating within a fluid, with x and z the horizontal and vertical directions, respectively. The force to the right is just the difference of the pressure forces between the right and left surfaces of the slab, and so proportional to δp. Thus, the net force is proportional to the pressure gradient within the fluid.

We may express the above arguments in mathematical form. Suppose now that the slab in figure 3.3 is just the fluid itself, with a cross-sectional area of A and a thickness of δx. The total pressure force on the left surface of the slab is the pressure times the area, equal to pA, and this force pushes the slab to the right (the positive x direction). The pressure on the right surface is equal to

$(p + \delta p)$, so the force pushing the slab to the left is equal to $(p + \delta p)A$. Thus, the net force pushing the slab to the right is equal to $-\delta pA$. The volume of the slab is just $A\delta x$, so that the net pressure force per unit volume is just $-\delta pA / A\delta x = -\delta p / \delta x$. That is, if we denote the pressure gradient force per unit volume in the x direction as F_x then we have

$$F_x = -\frac{\partial p}{\partial x}, \tag{3.5}$$

where we have used the notation of a partial derivative $(\partial p / \partial x)$ because pressure may be changing in other directions too. Thus, in words, the pressure gradient force per unit volume in a particular direction is equal to the rate of change of pressure with respect to distance in that same direction. Why is there a minus sign in equation 3.5? It is because the pressure force is directed from high pressure to low pressure. Thus, if the pressure is increasing in the x direction (i.e., to the right), so that $\partial p / \partial x$ is positive, then the pressure force is directed in the *negative x* direction, and to the left.

In the vertical direction, precisely the same considerations apply, so that the pressure gradient force in the vertical direction is given by

$$F_z = -\frac{\partial p}{\partial z}, \tag{3.6}$$

where z measures distance upward. If the pressure decreases upward (as it does), then $\partial p / \partial z$ is negative, and the pressure force F_z is postive and directed upward.

HYDROSTATIC BALANCE

Let us now consider what the balance of forces is in the vertical direction. A moment's thought indicates that there must be a pressure gradient in the vertical direction, because without one there would be nothing to hold up the fluid. A piece of fluid, be it a piece of air or a piece of seawater, is not weightless, so the force of gravity acts on it, pulling it down toward Earth. If the fluid parcel is stationary, then there must be a force in the opposite direction that balances gravity, and this is the pressure gradient force. That is to say, there is a balance between the gravitational force and the pressure gradient force, and this is known as *hydrostatic balance.* The balance holds exactly only when the fluid is not moving (hence the "static" in hydrostatic), but in Earth's atmosphere and ocean, it holds to a good approximation even when the fluid is moving.

To be quantitative, let us consider the vertical force balance on a thin slab of fluid of density (mass per unit volume) ρ, and let the slab have thickness δz and area A. The downward force due to gravity on the slab is just equal to g (the acceleration caused by gravity) times the mass on the fluid, $\rho A \delta z$. Thus, the total downward gravitational force per unit volume is $\rho g A \delta z$, or ρg per unit volume. The vertical pressure force on the slab is given by equation 3.6, and so we have

Downward gravitational force $= g\rho A \delta z$, \hfill (3.7)

$$\text{Upward pressure gradient force} = -A\delta z\frac{\partial p}{\partial z}. \qquad (3.8)$$

If the fluid is static, then the above two forces must balance, and we obtain

$$\frac{\partial p}{\partial z} = -\rho g. \qquad (3.9)$$

This is the equation of hydrostatic balance. (Again we use a partial derivative because pressure might also be changing in other directions.) The rate of change of pressure with height is thus proportional to the density of the fluid, and because seawater is about 1,000 times more dense than air, the pressure increases rapidly indeed as we go deeper into the ocean.

In the ocean, the density of seawater is almost constant and, if we neglect the relatively small contribution of atmospheric pressure at the surface of the ocean, equation 3.9 may be integrated from $z = 0$ (the ocean surface) to a depth d below the ocean surface to give

$$p = -\rho g z = \rho g d + p_0, \qquad (3.10)$$

where p_0 is the pressure at the ocean surface, which is small compared to p itself once d exceeds a few meters. (Note that we always measure z going up (that is, increasing in the upward direction) so $d = -z$.) The factor $\rho g d$ is just the weight of the water column per unit area from the surface to the depth d. Thus, to a good approximation, the pressure at any depth in the fluid is equal to the weight of the fluid (per unit area) above it.

This important result has many ramifications for ocean circulation.

GEOSTROPHIC BALANCE

Let us now consider what forces and balances occur in the horizontal direction, for these forces give rise to the ocean currents and the great circulation patterns in the ocean. Away from regions where the direct effects of wind and friction are important (usually at the top and bottom of the ocean), the two dominant forces in the horizontal direction are the pressure gradient force and the Coriolis force; if the flow is steady, these two forces almost balance each other. This balance is called *geostrophic balance*. The evolution of a flow into geostrophic balance may be envisioned as follows. Suppose that initially there is a pressure gradient in the fluid. The pressure gradient generates fluid flow from the high-pressure region to the low-pressure region, but as the fluid moves it is deflected by the Coriolis force. In the Northern Hemisphere, the fluid comes into equilibrium, with the Coriolis force trying to deflect the fluid to the right of its direction of motion and the pressure force trying to deflect it to the left. The direction of motion is perpendicular to both the pressure force and the Coriolis force.

Mathematically, we equate the Coriolis force given in equation 3.4 with the pressure gradient force of equation 3.5 and find

$$\text{Zonal direction:} \quad -fv = -\frac{1}{\rho}\frac{\partial p}{\partial x}, \quad (3.11)$$

Meridional direction: $\qquad fu = -\frac{1}{\rho}\frac{\partial p}{\partial y}.$ $\qquad$ (3.12)

These are the equations of geostrophic balance, and these equations hold to a good approximation for most large-scale flow in the ocean, especially away from boundaries.

How do horizontal pressure gradients arise in the ocean? Recall that the pressure at a point is primarily determined by the weight of the water above it. Thus, there is a horizontal gradient of pressure in the ocean interior either if the surface of the ocean is sloping or if the density of the seawater is varying. The latter effect may happen if the water has lateral gradients of temperature or salinity, as indeed is the case in the real ocean. We talk more about these effects in chapter 4, but let us perform a simple calculation to see how much slope of the ocean surface we need to produce a decent ocean current. The pressure at a point under the ocean is given by the weight of the column above it, so that $p = \rho g h$ where h is the height of the column. Using this replacement in equation 3.12 gives $fu = -g\partial h/\partial y$, where $\partial h/\partial y$ is the slope of the sea surface. Simple arithmetic then reveals that if the sea surface slopes by just 1 meter over a horizontal distance of 1,000 km, a geostrophic current of 10 cm s^{-1} is generated.

Finally, let's consider the question of why geostrophic balance should be dominant. Its dominance stems from the fact that the wind and currents are relatively weak compared to the speed of rotation of Earth itself. If the current changes by an amount U over a distance L, then the acceleration that the current undergoes is U^2/L.

In comparison, the Coriolis acceleration is fU, so that the ratio of the local acceleration to the Coriolis acceleration is U/fL. This ratio is called the *Rossby number*, after the Swedish meteorologist Carl-Gustav Rossby, and it is a very small number for currents in the ocean (the reader should calculate a representative value with $U = 0.1 \text{ m s}^{-1}$, $L = 1,000 \text{ km}$, and $f = 10^{-4} \text{ s}^{-1}$. Thus, the Coriolis effect is far larger than the local effect of the flow changing speed or direction, and so most flows in the ocean are, in fact, geostrophic flows satisfying equations 3.11 and 3.12.

EKMAN LAYERS

In the final topic of this chapter, we consider what happens to water in the upper few tens of meters when the wind blows over it. The water moves, and hence the Coriolis force acts, but because there is an additional force coming from the wind, geostrophic balance cannot exactly hold. So what does happen? The problem was first considered by Vagn Walfrid Ekman, a Swedish oceanographer, at the beginning of the twentieth century, at the suggestion of Fridtjof Nansen, the Norwegian explorer and statesman. Nansen had noticed that icebergs did not move in the same direction as the wind, but at an angle of about 45° to the right of the wind. In explaining this, Ekman was led to discover the eponymous Ekman layer, which we now discuss.

The blowing wind supplies a stress to the ocean and causes the uppermost layer of fluid to accelerate in the direction of the wind. As the fluid moves, two things happen. First, the fluid feels the effect of the Coriolis force, which in the Northern Hemisphere causes the fluid to veer to the right, as we have seen. Second, the fluid layer imparts a stress to the fluid just below the surface, and this layer is then set in motion. At the same time, the deeper layer provides a retarding force on the surface layer, so that the surface layer comes into a mechanical equilibrium, moving in a direction somewhat to the right of the wind at the surface. While this is happening, the deeper layer of fluid veers more to the right and imparts a stress to the still deeper fluid, and so on. The upshot of all this is that the flow veers more and more to the right with depth, and also becomes weaker with depth, as the wind's influence wanes. The net result is that the wind-induced flow forms a spiral, as illustrated in figure 3.4, with the flow magnitude typically falling essentially to zero after about 100 m, at which depth the wind-induced stress is negligible. Below this level, the flows are geostrophic.

In reality, such an ideal spiral is rarely, if ever, observed because of the myriad other processes occurring in the upper ocean. However, one robust property of Ekman layers transcends the fragility of the spiral structure itself. And that is that the mean transport in the Ekman layer—*Ekman transport*—is at right angles to the direction of the wind. The reason for this is relatively straightforward; it stems from the fact that the Coriolis force acts at right angles to the direction of the fluid flow. When the wind blows, it imparts a stress to the ocean,

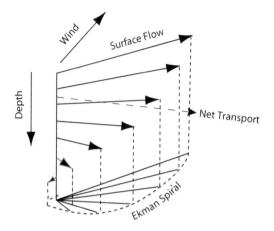

Figure 3.4. An idealized Ekman spiral in the Northern Hemisphere.

and the fluid begins to accelerate in the direction of the wind. The Coriolis force deflects the fluid (the deflection is to the right in the Northern Hemisphere), and the fluid comes into an equilibrium when the direction of the wind-induced flow is at right angles to the wind itself, for then the Coriolis force can exactly balance the wind stress. The situation is analogous to that giving rise to a geostrophically balanced flow, but now with the wind stress instead of a pressure gradient. Of course, in addition to the Ekman transport there is a geostrophic flow if there is a pressure gradient in the fluid. (Also, it turns out that just at the surface the flow is at 45° to the wind, as sketched in figure 3.4, thus explaining Nansen's observations of icebergs.) A more mathematical derivation of the Ekman transport is given in appendix B of this chapter, but trusting readers may be tempted to skip it.

APPENDIX A: DERIVATION OF CENTRIFUGAL AND CORIOLIS FORCES

In this appendix, we give an elementary mathematical derivation of the centrifugal and Coriolis forces that we encountered in this chapter.

The centrifugal force

In this section, we derive an expression for the centrifugal force on a body moving in a circle of radius r with an angular velocity Ω, as illustrated in figure 3.5.

Acceleration in a circular motion

Let us first consider what forces are involved in a body that is moving in a circle with a uniform speed. Newton's first law of motion says that if a body is left to its own devices it will either remain stationary or move in a straight line with a constant speed. If a body is undergoing uniform circular motion (for example, a child riding on a carousel or, more germanely for us, a person standing still on Earth and so rotating around Earth's axis once per day), then that body must be accelerating, with the acceleration directed toward the axis of rotation.

Let us suppose that in a small time, δt, the body moves through a small angle $\delta\theta$, so that the angular velocity is $\Omega = \delta\theta/\delta t$. The speed of the body is a constant,

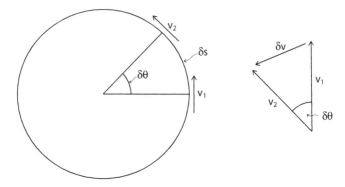

Figure 3.5. A body moving in a circle is constantly changing its direction and so accelerating. The acceleration is directed toward the center of the circle and has magnitude v^2/r, where v is the speed of the body and r is the radius of the circle.

$v = r\Omega = r\,\delta\theta/\delta t$, but its direction is changing such that at all times, the direction of motion is perpendicular to the radius; thus, if at the initial time the position vector is $\boldsymbol{r}_1$, then its velocity, $\boldsymbol{v}_1$, is in the perpendicular direction, and a short time later the velocity $\boldsymbol{v}_2$ is perpendicular to its new position vector, $\boldsymbol{r}_2$.

As can be seen in figure 3.5, the angle through which the radius vector has moved is related to the distance moved by

$$\delta\theta = \frac{\delta s}{r} \approx \frac{|\delta r|}{r}. \tag{3.13}$$

Because the velocity is always perpendicular to the radius, the direction of the velocity must move through the same angle as the radius and we have

$$\delta\theta = \frac{|\delta v|}{v}. \tag{3.14}$$

The acceleration of the body is equal to the rate of change of the velocity; using equation 3.14, we obtain

$$\frac{\delta|v|}{\delta t} = v\frac{\delta\theta}{\delta t} = \frac{v^2}{r}, \tag{3.15}$$

where the last equality uses the fact that $v = r\delta\theta/\delta t$. That is to say, the acceleration of a body undergoing uniform circular motion is directed along the radius of the circle and toward its center and has magnitude

$$a_{\text{circ}} = \frac{v^2}{r} = \Omega^2 r. \tag{3.16}$$

This kind of acceleration is called *centripetal acceleration*, and the associated force causing the acceleration (for there must be a force) is called the centripetal force and is directed inward, toward the axis of rotation.

Forces in the rotating frame of reference

Now let us consider the motion from the point of view of an observer undergoing uniform circular motion. An illuminating case to consider is that of a satellite or space station in orbit around Earth; when an astronaut goes for a space walk from the space station, she appears to be weightless, with no forces whatsoever acting upon her (figure 3.6). Now in fact, the astronaut is undergoing circular motion around Earth, and is therefore accelerating, and the force providing the acceleration is Earth's gravity.

a) Stationary or Inertial Frame

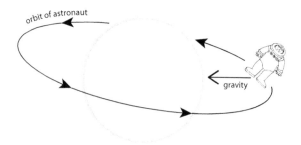

b) Rotating Frame of Reference

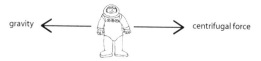

Figure 3.6. An astronaut orbiting Earth. Panel a views the motion in a stationary frame of reference, in which Earth's gravitational force provides the centripetal force that causes the astronaut to orbit Earth. Panel b views the situation from the astronaut's frame of reference, in which the gravitational force is exactly balanced by the centrifugal force and the astronaut feels weightless.

But from the point of view of an observer rotating with the astronaut, the astronaut is stationary and therefore no net forces are acting upon her. Now we know that gravity is acting, pulling her toward Earth, and we say that this force is balanced by another force, *centrifugal force*, which is pushing her out. The forces exactly balance, so the astronaut appears weightless; that is, in the rotating frame, the centripetal gravitational force pulling her toward Earth is exactly balanced by the centrifugal force pushing her out. Because the centripetal force has magnitude $\Omega^2 r$, the centrifugal force must have this magnitude also, and we conclude that in a rotating frame of reference, there appears to be an additional force, the centrifugal force, which acts to accelerate a body along a radius, outward from the axis of rotation. The magnitude of the centrifugal force is $\Omega^2 r$ per unit mass, where r is the distance from the axis of rotation.

There is no centrifugal force in inertial frames of reference. However, given that we live on a rotating planet, it is extremely convenient to describe motions on Earth from a rotating frame of reference in which Earth's surface is stationary, and in this frame it appears that there is a centrifugal force that is trying to fling us into space. Rather fortunately for those of us living on the planet's surface, the centrifugal force is much weaker than Earth's gravity.

Coriolis force

In this section, we give a mathematical, although elementary, derivation for the magnitude of the Coriolis force.

Coriolis force for a body moving zonally

Consider a disk rotating with an angular velocity Ω, so that at a radius r from the axis of rotation the disk has a tangential velocity $u_d = \Omega r$. Suppose that an observer is sitting on the disk and is thus stationary in the disk's frame of reference. As we saw in the previous section, in the rotating frame there is a centrifugal force of magnitude $\Omega^2 r$ trying to push the observer out, and in this case the centrifugal force is balanced by the friction between the observer and the disk.

Now let us suppose that the observer is passed by another body (a cyclist, for example) who is moving along a radius of the disk at a velocity u relative to the disk. The cyclist's total velocity, in the inertial frame, is $U = u_d + u = \Omega r + u$, and his total acceleration toward the axis of rotation, A_c, as measured in the inertial frame, is

$$A_c = \frac{U^2}{r} = \frac{(\Omega r + u)^2}{r} = \Omega^2 r + \frac{u^2}{r} + 2\Omega u. \tag{3.17}$$

This acceleration must be caused by a centripetal force directed toward the axis of rotation, equal to the mass of the cyclist (M, say) times his acceleration, and this acceleration is provided by the friction of the wheels against the surface and by the cyclist leaning in. That is, the inward force is given by $F_c = MA_c = M(\Omega^2 r + u^2/r + 2\Omega u)$, where M is the mass of the cyclist.

Let us now consider how the forces and accelerations appear to the observer sitting in the rotating frame. The inward force on the cyclist is still F_c because the frictional force and the force due to the cyclist leaning in are still present. There is also an outward centrifugal force on the cyclist equal to $M\Omega^2 r$ because we are measuring everything in the rotating frame. Thus, the net apparent inward force on the cyclist, as measured by the observer, is $F_c - M\Omega^2 r = M(u^2/r + 2\Omega u)$.

Regarding the acceleration, the observer measures the cyclist to be moving in a circle with a speed u, and therefore with an apparent acceleration toward the center of the disk of just u^2/r. Thus, in the rotating frame, even after properly taking into account the centrifugal force, there appears to be a mismatch between the inward forces per unit mass ($u^2/r + 2\Omega u$) and the inward acceleration u^2/r. We reconcile ourselves to this mismatch by saying that, in the rotating frame, there is an additional force on a moving body equal to $2\Omega u$ per unit mass. This force is the Coriolis force, and it acts at right angles to a body moving in a rotating frame of reference.

To summarize, in the inertial frame there is an inward centripetal force on the cyclist given by

$$F_c = M(\Omega^2 r + \frac{u^2}{r} + 2\Omega u) \tag{3.18}$$

and a consequent inward acceleration given by

$$A_c = \Omega^2 r + \frac{u^2}{r} + 2\Omega u. \tag{3.19}$$

In the rotating frame, there is an inward acceleration given by

$$A_c^{\text{rot}} = \frac{u^2}{r}. \tag{3.20}$$

The inward force is caused by the real frictional and leaning forces on the cyclist, F_c, as given by equation 3.18, but this force is greater than that required to produce a balance in the rotating frame. Such a balance is achieved by positing outward centrifugal and Coriolis forces:

$$F_{\text{cent}} = M\Omega^2 r, \qquad F_{\text{Cor}} = 2M\Omega u. \qquad (3.21\text{a, b})$$

Coriolis force for a body moving meridionally

We now consider the Coriolis force for a body moving in the meridional direction, that is, along a line of longitude. As in the previous section, we consider flow on a rotating disk for which meridional flow corresponds to radial flow.

Suppose a body is sitting on the rotating disk and so is stationary in the rotating frame of reference, held in place by friction against the centrifugal forces pushing it out. Let us now suppose that we push it toward the axis of rotation. One of the fundamental consequences of Newton's laws of motion is that, unless *tangential* forces are acting on the body, its angular momentum, m, around the axis of rotation is conserved. We express this mathematically as

$$m \equiv Ur = (\Omega r + u)r = \Omega r^2 + ur = \text{constant}, \qquad (3.22)$$

where Ω is the angular velocity (proportional to the rate of rotation) of the disk, r is the distance from the axis of rotation, and u is the velocity in the tangential direction. The quantity $U = \Omega r + u$ is just the tangential velocity in the absolute (or nonrotating) frame of reference. The initial velocity is in the radial direction, so the tangential velocity is initially zero, but we may anticipate that the Coriolis force will deflect the body in the tangential direction.

Let us differentiate equation 3.22 with respect to time:

$$\frac{\mathrm{d}m}{\mathrm{d}t} = \Omega \frac{\mathrm{d}r^2}{\mathrm{d}t} + \frac{\mathrm{d}u}{\mathrm{d}t}r + u\frac{\mathrm{d}r}{\mathrm{d}t} = \Omega 2r\frac{\mathrm{d}r}{\mathrm{d}t} + \frac{\mathrm{d}u}{\mathrm{d}t}r + u\frac{\mathrm{d}r}{\mathrm{d}t}. \qquad (3.23)$$

Now, because m is constant, $dm/dt = 0$, but both r and u may change. The derivative of r with respect to time is just v, the velocity in the radial direction, so that equation 3.23 becomes

$$0 = 2\Omega v r + \frac{\mathrm{d}u}{\mathrm{d}t}r + uv \qquad (3.24)$$

or, rearranging the terms,

$$\frac{\mathrm{d}u}{\mathrm{d}t} = -2\Omega v - \frac{uv}{r}. \qquad (3.25)$$

The term on the left-hand side is the apparent acceleration of the body, as seen in the frame of reference of the rotating disk. The right-hand side must then be equal to the forces acting on the body, divided by the mass of the body. How do we interpret these forces? The first term is just the Coriolis force caused by the rotation of the disk itself; it is our old friend $2\Omega v$, with the minus sign just giving us the direction of the force. If a flow is toward the axis of rotation, then $v = \mathrm{d}r/\mathrm{d}t$ is negative, which produces a force pushing the body to the right, with a positive u. The second term on the right-hand side, uv/r is usually much smaller than the Coriolis term and is another apparent force that arises because of the movement of the body itself. (It is sometimes called a metric term, as it only appears in non-Cartesian coordinates.)

We have thus shown that when a body moves in the radial direction it experiences a Coriolis force, equal to twice the angular velocity multiplied by the velocity in the radial direction. This force deflects the body in the tangential direction, perpendicular to the direction in which the body is moving. We saw earlier in this chapter that when a body is moving in the tangential direction it experiences an apparent force in the *radial* direction, now equal to twice the rotation rate multiplied by the velocity in the tangential direction. Now, the velocity in an arbitrary direction can always be decomposed into a velocity in the radial direction plus a velocity in the tangential direction, and we may conclude that, in general, a body moving in a rotating frame experiences a force at right angles to the direction of its velocity, and of magnitude equal to twice the rotation rate times the speed of the body. This force is the Coriolis force, and it deflects bodies to the right in the Northern Hemisphere and to the left in the Southern Hemisphere.

APPENDIX B: FLOW IN AN EKMAN LAYER

Most large-scale flow in the ocean is in geostrophic balance, meaning that, away from the direct influence of the wind and so beneath the Ekman layer, the pressure gradient force is balanced by the Coriolis force. Mathematically we have, as in equation 3.11,

$$-f\rho v_g = -\frac{\partial p}{\partial x}, \quad f\rho u_g = -\frac{\partial p}{\partial y}, \quad \text{(3.26a, b)}$$

where u_g and v_g are the fluid speeds in the zonal and meridional directions, respectively. The first of this pair of equations is the momentum balance in the zonal direction (with x the distance toward the east): the Coriolis force $-fv_g$ balances the pressure gradient force. Similarly, the second equation is the momentum balance in the meridional (y) direction. These equations may be regarded as *defining* the geostrophic velocities u_g and v_g (note the subscript g on the variables).

Now suppose that we add a stress, τ, to this balance. The stress is provided by the wind, and it diminishes with depth. The force is the vertical derivative of the stress so that equations 3.26a and b become

$$-f\rho v = -\frac{\partial p}{\partial x} + \frac{\partial \tau^x}{\partial z}, \quad f\rho u = -\frac{\partial p}{\partial y} + \frac{\partial \tau^y}{\partial z}, \qquad \text{(3.27a, b)}$$

where u and v are the zonal and meridional components of the total velocity (not just the geostrophic flow) in the upper ocean, and τ^x and τ^y are the corresponding components of the stress. We rewrite equations 3.27a and b as

$$-f\rho(v - v_g) = \frac{\partial \tau^x}{\partial z}, \quad f\rho(u - u_g) = \frac{\partial \tau^y}{\partial z}, \qquad \text{(3.28a, b)}$$

having used equations 3.26a and b as the definitions of geostrophic velocity. If we integrate equations 3.28a and b over the depth of the Ekman layer (i.e., from the surface to the depth at which the stress vanishes), we obtain

$$-fV = \tau^x_w, \quad fU = \tau^y_w, \qquad \text{(3.29a, b)}$$

where τ^x_w and τ^y_w are the zonal and meridional components of the wind stress at the surface, and $V = \int \rho(v - v_g)$ dz and $U = \int \rho(u - u_g)$dz are the meridional and zonal components of the wind-induced, nongeostrophic mass transports, integrated over the Ekman layer. Thus, we see that a *zonal* wind stress (i.e., a nonzero τ_w^x) induces a *meridional* flow in the ocean, and a meridional wind stress induces a zonal flow. That is, the average induced velocity in the Ekman layer—the *Ekman transport*—is perpendicular to the imposed wind stress at the surface. In the Northern Hemisphere where f is positive, if the stress is eastward (i.e., τ_w^x is positive), then the Ekman transport is southward (V is negative).

4 THE OCEAN CIRCULATION

> This is a court of law, not a court of justice.
>
> —Oliver Wendell Holmes

THE CLIMATE IN GENERAL AND THE OCEANS IN PARTIC-
ular are complicated systems, and if one is not careful it
is easy to lose sight of the forest for the trees. For that reason, a useful philosophy is to begin with an austere picture of the phenomenon at hand and then gradually add
layers of complexity and detail. The first picture will be a
simplification, but if it is based on sound scientific principles, then it will provide a solid foundation for what
follows, and it will become possible to work toward an
understanding of the system as it really is. In this chapter we apply this philosophy to try to understand the
ocean circulation. We won't seek a full understanding of
the real system; rather, we will construct a physical and
mathematical representation of it, a model based on the
same laws of physics that are satisfied by the real ocean.

WHAT MAKES THE OCEAN CIRCULATE?

As we discussed in chapter 2, it is useful to think of the large-
scale ocean circulation as having two main components:

a quasi-horizontal circulation consisting of the gyres and other surface-enhanced currents, and a deeper overturning circulation, the meridional overturning circulation. What makes the ocean go around this way? What "drives" the ocean, if anything? Bypassing the ambiguous term "drive," there are three main distinct physical phenomena that lead to the circulation of the ocean:[1]

1. The mechanical force of the wind on the surface of the ocean provides a stress that produces a quasi-horizontal circulation that includes, most noticeably, the *wind-driven gyres*. The predominantly horizontal currents of the world's ocean, shown in figure 2.3 in chapter 2, are primarily a consequence of wind forcing. Less obviously, the wind also plays a role in producing a deep, interhemispheric meridional overturning circulation, a circulation in which the water sinks near one pole and rises near the other.

2. Buoyancy effects, caused mainly by the cooling of the oceans at high latitudes and heating at low latitudes, generally produce denser water at high latitudes. Salinity is a secondary source of density gradients in today's climate. An overturning circulation arises in response to these density gradients with cool, dense water sinking at high latitudes, moving equatorward and rising at lower latitudes and/or in the opposite hemisphere.

3. The mixing of fluid properties, and in particular heat, by small-scale turbulent motions (sometimes called turbulent diffusion) brings heat down into the abyss and enables an overturning circulation to be maintained.

The gyres and other quasi-horizontal currents are mainly a response to winds, and although they are affected by buoyancy effects and mixing, we can safely call them wind driven. The meridional overturning circulation (MOC), on the other hand, involves all three effects in an essential way. Most obviously, the MOC arises as a response to the surface density gradients (item 2 in our list) and is sometimes called the *thermohaline circulation*, so-called because it is enabled by the buoyancy effects of heat and salt leading to the sinking of dense water. However, we will see that the MOC can only be *maintained* if either mixing or wind is present, for they enable the deep water to rise to the surface to begin circulating anew. Without them, the deep circulation would stagnate.

Let's first discuss the wind-driven circulation, the great gyres, and western intensification, and follow that with a discussion of the MOC. The equatorial currents are different again, and we defer discussing them until chapter 6.

THE WIND-DRIVEN CIRCULATION AND THE GREAT GYRES

To better understand how the processes described above produce an ocean circulation like that described in chapter 2, let us consider an idealized ocean, with much simplified geometry, and see if we can first understand how that works. Our idealized view of the gyres is illustrated in figure 4.1, which the reader will appreciate is an abstraction

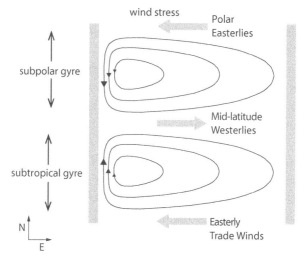

Figure 4.1. An idealized gyre circulation in a rectangular ocean basin in the Northern Hemisphere, showing the subtropical gyre (lower, typically extending from about 15°N to 45°N), the subpolar gyre (upper), and the intense western boundary currents on the left.

of the real circulation of the world's ocean. The main questions we wish to answer are relatively simple:

1. Why do the gyres exist in the first place? What determines the way they go around and how strong they are?
2. Why are they more intense on the western sides of the oceans?

The gyres exist because the mean winds provide a mechanical forcing, a stress, on the oceans, and this stress causes the water to accelerate. For the oceans to be in mechanical balance, the imposed forces must be counteracted by frictional forces where the water rubs against the ocean bottom or side. Frictional forces only arise when the water is in motion, so that if there is a wind blowing, then the ocean must be in motion, and an overall balance between the wind and the frictional forces ultimately comes about. However, there are important effects caused by Earth's rotation that determine the structure of the gyres, as we will see.

For the sake of definiteness, we consider the subtropical gyre in a rectangular ocean—the lower gyre of figure 4.1. The winds blow eastward on the poleward side of the gyre (these are the midlatitude westerly winds) and westward at low latitude (the tropical trade winds), and it seems entirely reasonable that the ocean should respond by circulating in the manner shown. However, in the last chapter we noted that Earth's rotation plays a significant role in large-scale circulation and that flows are generally in geostrophic balance, except for the Ekman layer in the upper ocean, where the flow is at right angles to the wind. How does this description square with the notion of a gyre that seems to go around in the same direction as the wind?

The Ekman and wind-induced geostrophic flows

We show first that the wind does indeed induce a geostrophic flow that has the same sense as the wind itself. The mean winds are to the east in midlatitudes and to the west in the tropics and, as we showed in the section in chapter 3 on Ekman layers, there is a flow in the upper

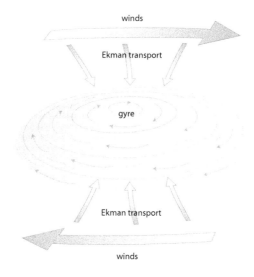

Figure 4.2. Production of gyres by winds. The winds blowing as shown induce a converging Ekman flow, causing the sea level to increase in the center, thus giving rise to a pressure gradient. This gradient in turn induces a geostrophic flow around the gyre, in the same sense as the winds themselves.

ocean at right angles to the wind. As illustrated in figure 4.2, this combination causes the flow to converge in the center of the gyre. This convergence pushes up the surface of the ocean, causing the sea surface to form a gentle dome, with the ocean surface at the center of the gyre a few tens of centimeters higher than at the edges. The converging fluid must go somewhere, and the only place for it to go is downward. A complementary situation arises in the subpolar gyre, where the westerly (eastward) winds are strongest on the equatorial side. Now the Ekman transport is directed *away* from the center of the gyre, and the sea level is depressed and upwelling occurs.

The doming of the sea surface produces a pressure gradient in the ocean, as illustrated in figure 4.2. Consider a horizontal plane at a level a little below the sea surface. The pressure at that level is produced by the weight of the fluid above it, as we discovered in the section in chapter 3 on hydrostatic balance, and so is higher where the sea surface is higher. This pressure gradient produces a geostrophic flow perpendicular to the pressure gradient, and so in the same direction as the wind that originally produced the doming. Thus, when all is said and done, on a rotating planet the wind leads to the production of an ocean current that is aligned with the wind, rather as we would expect in the nonrotating case. However, the pressure gradients in the two cases are quite different because of the presence of the Coriolis force in the rotating case; note in particular that the horizontal pressure gradient produced by the doming extends all the way to the bottom of the ocean. Thus, even though the direct effects of the wind stress are confined to the upper few tens of meters, the wind produces geostrophic currents that can extend to great depths.

Sea-surface slope and the geostrophic current

It may seem a little fantastical that the wind can produce a change in the sea level and that this in turn can produce the currents of the great gyres. However, we do not need a large change in the sea level to produce quite substantial flows, as we can see with a simple calculation. The geostrophic current is a balance between the Coriolis and pressure gradient forces, so that

$$fu = -\frac{1}{\rho}\frac{\partial p}{\partial y}, \qquad -fv = -\frac{1}{\rho}\frac{\partial p}{\partial x}. \tag{4.1a, b}$$

The pressure at a level below the surface is given by the weight of the fluid above it, so that

$$p = \rho gh, \tag{4.2}$$

where h is the height of the column of seawater and ρ is the density of the seawater. Thus, using equation 4.2 in equation 4.1a and b, we get

$$fu = -g\frac{\partial h}{\partial y}, \qquad fv = g\frac{\partial h}{\partial x}. \tag{4.3a, b}$$

Suppose that the height of the sea surface varies by just 1 m over a horizontal distance of 1,000 km—a truly small slope that would be extremely difficult to detect by measurements of the ocean surface but that is, remarkably, measurable using modern satellites. The magnitude of the currents produced is then given by

$$u = \frac{g}{f}\frac{\Delta h}{L} = \frac{9.8}{10^{-4}}\frac{1}{10^{6}} \approx 0.1 \text{ m s}^{-1}. \tag{4.4}$$

Such a current is easily measurable, and when one considers that billions of tons of water might be put in motion this way, one begins to see the large effect that this current can have.

WESTERN INTENSIFICATION

We have now explained the underlying reason for gyres, but we have not explained one of their most important and indeed obvious aspects: the gyres are not symmetric in the east–west direction. Thus far, our explanation would lead to gyres that look like those in the left panel of figure 4.3, whereas in fact the gyres look more like those in the right-hand panel, with intense *western boundary currents,* of which the Gulf Stream in the western North Atlantic is the most famous example to Americans and Europeans, and the Kuroshio is the corresponding current off the coast of Japan. The presence of the Gulf Stream has been known for a long time—Benjamin Franklin was one of the first people to chart it. So our question is a simple one: why is the Gulf Stream in the west?

It turns out that the cause of the western intensification is that, as we discussed in chapter 3, the magnitude of the effective rate of rotation (specifically, the magnitude

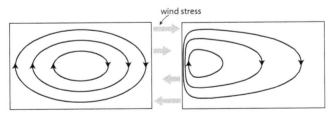

Figure 4.3. Two schematics of a subtropical gyre. The left panel shows the basic response of the circulation to the winds shown, and the right panel shows the gyres in the presence of differential rotation, with western intensification.

of the Coriolis parameter) of Earth increases as we go poleward. This *differential rotation* causes the gyres to have a marked east–west asymmetry, with the flow in the west squished up against the coast. As the effect is both important and hard to grasp, we give a couple of explications. For definiteness we focus on the subtropical gyre in the Northern Hemisphere, but the same principles apply to the other gyres.

Torques and interior flow

If the wind stress acting on the ocean varies with latitude—as we see that it does in figure 4.3—then the wind provides a torque that tends to *spin* the ocean. In a steady state, not only do the forces on the ocean have to balance but so do the torques; otherwise the ocean would spin faster and faster. The torques on the ocean are provided by the wind, by friction, and by the Coriolis force (the pressure gradient does not provide a torque).[2] Integrated over the entire ocean basin, the wind torque is balanced by the frictional torque, and since the frictional torque normally acts in a sense opposite to that of the spin of the fluid itself, the basin-scale circulation spins in the same sense as the wind. That is, for there to be a balance between wind and friction, the large-scale flow must have the same overall sense of rotation as the wind, producing the gyre shown in the left panel of figure 4.3.

However, in the interior of the basin, frictional effects are in fact very weak and the spin provided by the wind stress is locally balanced by the effects of the Coriolis force. Now, the Coriolis parameter increases northward, and it turns out that to locally balance the wind torque, a meridional flow must be produced in the ocean interior. The direction of the meridional flow depends on the sense of the spin provided by the wind, but in the subtropical gyre the meridional flow turns out to be equatorward. Let's see why.

Consider a parcel of fluid in the middle of the ocean, as illustrated in figure 4.4. The wind blows zonally with a stronger eastward wind to the north and so provides a clockwise torque. We can balance this torque by a Coriolis force if there is a *southward* flow of water in the interior. In that case, the Coriolis force provides a westward force on all the parcels of fluid moving south. However, the force is stronger on the fluid that is in the northern part of the domain (because the Coriolis parameter increases northward), so the spin provided to the fluid is counterclockwise, opposing that spin provided by the wind. The southward flow adjusts itself so that the spin provided by the varying Coriolis force just balances the spin provided by the wind. (The balance is called *Sverdrup balance*, and the southward flow is called the Sverdrup interior.)

Now, the southward-flowing water must return northward somewhere, and this return must be at either the eastern or western boundaries because here the frictional effects of the water rubbing against the continental shelf and coast are potentially able to allow the flow to achieve a torque balance and move northward. However, only if the boundary layer is in the west (as illustrated in the right panel of figure 4.3) can such a balance be achieved: the

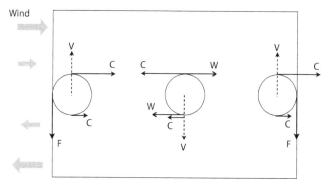

Figure 4.4. The production of a western boundary current. Schematic of the torques (namely, the spin-inducing forces: the wind, W; Coriolis, C; and friction, F) acting on parcels of water in the ocean interior (center) and western and eastern boundary layers (left and right), in a Northern Hemisphere subtropical gyre. In the interior, friction is small and the torques balance if the flow (denoted V) is southward. If the northward return flow is in the west, then a balance can be achieved between friction and Coriolis forces, as shown. If the northward return flow is in the east, no balance can be achieved.

gyres then circulate in the same sense as the wind forcing, and the frictional forces at the western boundary act to oppose the wind forcing and achieve an overall balance. If the return flow were to be in the east, then the flow would, perversely, be circulating in the opposite sense to the torque provided by the wind, and no balance could be achieved. A local view of how the torque balances work in the boundary layer is provided in figure 4.4.

Suppose that the wind blew the opposite way. The balance of the wind torque and the Coriolis effect can now be achieved if interior flow is northward, and this is the case in the subpolar gyres. For the overall flow to have the same sense as the wind torque, the return flow still has to be in the west. Thus, we see that western boundary currents are a consequence of the differential rotation of Earth, not the way the wind blows. If Earth rotated in the opposite direction, the boundary currents would be in the east.

Westward drift

In this section, we give a slightly different explication of why the boundary current is in the west. It is not really a different explanation because the cause is still differential rotation, but here we think about it quite differently. We'll see that the effect of differential rotation is to make patterns propagate to the west, and hence the response to the wind's forcing piles us in the west and produces a boundary current there.

We noted already that the component of Earth's rotation in the local vertical direction also increases as we move northward or, putting it a little informally, the *spin* increases northward. (The spin is also called the *vorticity*.) Now consider a parcel of fluid sitting in the ocean. It may be spinning from two causes, namely, because it is spinning relative to Earth and because Earth itself is spinning. If that parcel moves and if no external forces act upon it, then the total spin of the fluid parcel is preserved. Its local spin relative to Earth must therefore change to compensate for changes in Earth's spin.

Let's now imagine a line of parcels, as illustrated in figure 4.5. Suppose we displace parcel A northward.

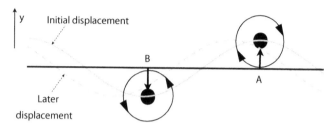

Figure 4.5. If parcel A is displaced northward, then its clockwise spin increases, causing the northward displacement of parcels that are to the west of A. A similar phenomenon occurs if parcel B is displaced south. Thus, the initial pattern of displacement propagates westward.

Because Earth's spin is counterclockwise (looking down on the North Pole) and this spin increases as the parcel moves northward, then the parcel must spin more in a clockwise direction to preserve its total spin. This spin has the effect of moving the fluid that is just to the west of the original parcel northward, and then this fluid spins more clockwise, moving the fluid to its left northward, and so on. The northward displacement thus propagates westward, whereas parcels to the east of the original displacement are returned to their original position so that there is no systematic propagation to the east. Similarly, a parcel that is displaced southward (parcel B) also causes the pattern to move westward. This is an idealized example—in fact we have just described the westward propagation of a simple *Rossby wave*—but the same effect occurs with more complex patterns and in particular, with the gyre as a whole. Thus, imagine that an east–west symmetric gyre is set up, as in the left panel of figure 4.3, with the winds and friction in equilibrium. Differential rotation then tries to move the pattern westward, but of course the entire pattern cannot move to the west because there is a coastline in the way! The gyre thus squishes up against the western boundary in the manner illustrated in the right panel figure 4.3, creating an intense western boundary current. This way of viewing the matter serves to emphasize that it is not the frictional effects that cause western intensification; rather, frictional effects allow the flow to come into equilibrium with an intense western boundary current, with the ultimate cause being the westward propagation caused by differential rotation.

THE OVERTURNING CIRCULATION

The other main component of the ocean circulation is the *meridional overturning circulation* (MOC), circulation essentially occurring in the meridional plane. There are two rather distinct aspects to this circulation, but they each have a common feature, namely the sinking of dense water at high latitudes and its subsequent rise to the surface elsewhere. Thus, in general the overturning circulation may be regarded as being "buoyancy enabled" in the sense that without buoyancy gradients at the surface there would be no deep overturning circulation. The buoyancy gradients themselves are produced by variations in temperature and salinity, and so the circulation is also sometimes known as the thermohaline circulation. The two different aspects are the processes that keep the water circulating. In one case, it is mixing by small-scale turbulent motions, and in the other case, it is the direct effect of the wind. We'll deal with these in turn, but before that, we discuss the buoyancy force itself.

The buoyancy or Archimedean force

The force due to buoyancy is one of the most familiar forces occurring in a fluid and, rather famously, was known to Archimedes. It is the force that, among other things, allows objects to flow in water. The Archimedes principle is often stated as "Any object, partially or wholly immersed in a fluid, experiences an upward force equal to the weight of the fluid displaced by the object." Let's see why this is so.

Consider a container of still water and focus attention on a particular piece of water that is fully surrounded by other fluid. The parcel has a finite weight, of course, and it does not sink to the bottom of the container because it is held up by the pressure force provided by the rest of the fluid in the container. Because none of the water is moving, the weight of the parcel (its mass times the acceleration due to gravity, acting downward) must exactly equal the upward pressure forces provided by the rest of the fluid. Now, let us replace the parcel with a solid object of the same shape and size. The upward pressure force provided by the rest of the fluid remains the same; this, we just ascertained, is equal to the weight of the parcel of fluid displaced—and this is Archimedes' principle. If the solid object is lighter than the weight of the fluid displaced, then there is a net upward force on it, and the object moves upward until it floats on the surface. If the solid object is heavier than the fluid displaced, the object sinks. These considerations apply to water itself. If we cool the water at the surface of the ocean, or add salt to it, it becomes more dense and therefore sinks—and it can sink quite quickly. A parcel of water that is negatively buoyant at the surface of the polar ocean can sink to considerable depth in a concentrated convective plume in a matter of hours to days, with a corresponding vertical velocity of a few centimeters per second. Similarly, if we warm the water that is at the bottom of the ocean, it will become lighter and rise, although this tends to be a much slower process, spread out over a wide area.

The overturning circulation maintained by mixing

How do the above considerations apply to the circulation of the ocean? For simplicity, we consider only the effects of temperature and not of salinity, and a schema of the circulation is given in the top panel of figure 4.6. The ocean, is, roughly speaking, a big basin of water for which the temperature of air just above the sea surface decreases with latitude. Air–sea exchange of heat heats or cools the water at the sea surface so that it has, approximately and on average, the temperature of the air above it. The sea-surface temperature thus decreases more or less monotonically from the equator to the pole, and as a consequence the density of the water at the sea surface increases from the equator to the pole.

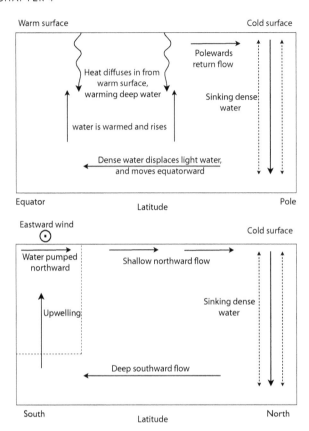

Figure 4.6. Schema of the two main components of the MOC. Top: The mixing-maintained circulation. Dense water at high latitudes sinks and moves equatorward, displacing warmer, lighter water. The cold, deep water is slowly warmed by diffusive heat transfer (mixing) from the surface in mid- and low latitudes, enabling it to rise and maintain a circulation. Bottom: Winds over the Antarctic Circumpolar Current (outlined by dashed lines) pump water northward, and this pumping enables deep water to rise and maintain the circulation. In the absence of both wind and mixing, the abyss would fill up with the densest available water and the circulation would cease.

As we mentioned, a fluid parcel itself sinks if it is cold and sufficiently dense. This is just what happens to water at high latitudes, especially in winter in the North Atlantic and near Antarctica, and this process is known as convection. Some lighter water at depth comes up to the surface to take the place of the dense, sinking water, as indicated by the dashed lines in figure 4.6, and as this water comes into contact with the cold atmosphere, it too cools and sinks, so that eventually the whole column of water at high latitudes is cold and dense. What happens then? Recall that the pressure at some level in a fluid is equal to the weight of the fluid above that level, so that if a column of fluid is cold and therefore dense, then the column weighs more than does a column of lighter fluid. Thus, the pressure in the deep ocean is largest at the high latitudes because the cold water weighs more than the warmer water at low latitudes. Thus, in the deep ocean there is a pressure force acting to push fluid from high latitudes to low latitudes, and the water begins to circulate, flowing at depth from high latitudes to low latitudes.

If no other physical processes occurred, the dense water would displace light water until the entire deep ocean were filled up with cold, dense water with polar origins. Nearer the surface, there would be a region of strong vertical temperature gradients, linking the low temperature of the abyss with the warmer surface waters. However, the deep, abyssal waters would eventually stop circulating because the water in the deep ocean would be as cold and dense as the coldest and densest waters at high latitudes at the surface. That is, the surface water would no longer be denser than the water beneath it, and convection and the deep circulation would cease. This state would be the "cold death" of the ocean.

So what enables a deep circulation to continue? The circulation continues because the deep water in low and midlatitudes is continually, albeit weakly, warmed by the transport of heat from the surface. This warming enables the water to rise and the circulation to continue. If there were no such heat transport, the deep ocean would simply fill up with cold, dense polar water. There would then be no convection because the cold surface waters at high latitudes would not be negatively buoyant. Thus, although the circulation can be thought of as being set up by a buoyancy gradient at the surface, its continuation relies on the effect of transport of heat down into the abyss, and without that, this part of the overturning circulation could not be maintained.

What physical process causes the downward heat transfer? In a *quiescent* fluid, the heat is transferred by molecular diffusion, in which molecules of water pass on their energy to neighboring molecules without any wholesale transport of fluid itself. However, the molecular diffusivity is very small and molecular diffusion is a slow process indeed, requiring thousands of years for a significant amount of heat to be diffused from the surface to the abyss. In fact, the ocean is a turbulent fluid, and the downward transport of heat is mainly effected by small-scale turbulent eddies. This process is sometimes called turbulent diffusion because the process is similar to that of molecular diffusion but with parcels of water replacing individual molecules. (Turbulent diffusion arises in large part from internal gravity waves that break and mix the fluid. Such waves, analogous to waves on the surface of the ocean but interior to the fluid, are generated by mechanical forcing—by the winds and the tides. Thus, without the effects of mechanical forcing, this component of the MOC would be weak indeed because the diffusion would be small.) Thus, to summarize, the following two effects combine to give an overturning circulation.

1. A meridional buoyancy gradient between the equator and the pole enables dense water to form at the surface at high latitudes and then potentially to sink in convective plumes and move equatorward. In today's ocean, the buoyancy gradient is predominantly produced by the temperature gradient.
2. The slow warming of the abyssal waters by turbulent diffusion of heat from the surface in mid- and low latitudes imparts buoyancy to the deep water and enables it to rise. Without this warming, the abyss would fill with cold, dense water and circulation would cease.

It is natural to think of the meridional buoyancy gradient as being between the equator and the pole, mainly caused by temperature falling with latitude. In this case, we can envision a meridional circulation in each hemisphere, with sinking at each pole and rising motion in mid- and low latitudes, in both hemispheres.

If one hemisphere were to be significantly colder than the other, then the abyss in both hemispheres could be expected to fill up with the water from the colder and denser hemisphere, which would create an interhemispheric circulation (more on that later). Finally, although we've couched our description in terms of the temperature effects on buoyancy, the effects of salinity can also be important. Salty water is heavier than freshwater at the same temperature, so adding salt can have a similar effect to that of cooling the surface. In today's climate, temperature has a larger effect than salinity on the variations in buoyancy so that the circulation is thermally driven, rather than salt driven. However, variations in salinity turn out to be the key difference in the overturning circulation of the Atlantic and the Pacific—the North Atlantic is saltier than the North Pacific, and so it can more easily maintain an overturning circulation.

The overturning circulation maintained by wind

The second mechanism that can lead to a deep overturning circulation relies, in its simplest form, on the presence of strong zonal wind blowing over the ocean surrounding Antarctica, as illustrated in the lower panel of figure 4.6. Unlike an ocean basin, the ocean surrounding Antarctica is effectively a channel, for it has no meridional boundaries and so no real gyres. Let's first look at the flow in this channel, and then look at how this flow affects the global overturning circulation. The wind around Antarctica blows in a predominantly zonal direction, toward the east. As one might expect, the wind generates a mean current in the same direction—the Antarctic Circumpolar Current, or ACC. However, because Earth is rotating, the wind stress generates an Ekman flux (as described in chapter 3) that is perpendicular to the wind, and so northward (the Coriolis force deflects bodies to the left in the Southern Hemisphere), as illustrated in figure 4.7.

The northward-flowing water in the Ekman layer must be compensated by southward-moving water to maintain a mass balance. In a gyre, the return flow could be at the surface in a western boundary current, but none exist in the ACC and the flow must therefore return at depth, where friction along the bottom enables the flow to be nongeostrophic, or the presence of topography allows zonal pressure gradients to be maintained. Where does the deep water ultimately come from? One option would be that the flow simply circulates locally in the Southern Hemisphere. However, if the water in the Northern Hemisphere is sufficiently dense, then it will be drawn into the Southern Hemisphere and into and across the ACC, where it can come up to the surface. Water at high latitudes in the North Atlantic is in fact sufficiently dense for this to occur, although water in the North Pacific is not (the key difference is salinity—the North Atlantic is saltier than the North Pacific). Thus, the presence of winds in the Southern Ocean generates an interhemispheric meridional overturning circulation, in which water sinks at high northern latitudes

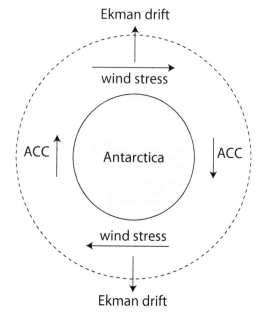

Figure 4.7. Schematic of the flow in the Antarctic Circumpolar Current (ACC). The wind predominantly blows in a zonal direction around the Antarctic continent, generating an Ekman flow toward the north and a net loss of water from the channel. The water returns at depth, generating a deep overturning circulation, as illustrated in the bottom panel of figure 4.6.

and moves southward across the equator, upwelling in the ACC. Unlike the mixing-maintained circulation described in the previous section, no mixing is required to draw up the deep water; rather, the wind itself pumps the deep water up.

Putting it all together

Thus, to summarize, the meridional overturning circulation has two mechanistically distinct components:
a component maintained by mixing and a component
maintained by wind, both responding to the surface
buoyancy distribution. The two can exist side by side,
and the overturning circulation in the Atlantic Ocean is
schematically illustrated in figure 4.8. Some of the water
that sinks in the North Atlantic moves across into the
Southern Hemisphere and upwells in the ACC (enabled
by the wind), and some upwells and returns in the North
Atlantic itself (enabled by mixing). The water that sinks
in the North Atlantic (forming the North Atlantic Deep
Water) does not in fact extend all the way to the bottom
of the ocean because there is some even denser water
beneath it—Antarctic Bottom Water, which comes from
high southern latitudes and circulates through the effects
of mixing.

Which component of the circulation is dominant?
Only careful observations can tell us, although currently
it is often believed that the wind component is stronger
than the mixing component in the Atlantic Ocean. The
North Pacific Ocean is generally less dense than the North
Atlantic because it is fresher; also it does not support a vigorous interhemispheric circulation and so partakes more
weakly in the global-scale overturning circulation that is
sketched in figure 2.6. Note finally that the horizontal velocities in the abyssal ocean are usually quite small, on the
order of 1 mm s^{-1}, and at this speed it would take some 300

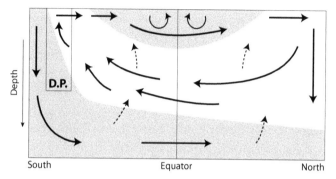

South   Equator   North

Figure 4.8. Schematic of the meridional overturning circulation, most applicable to the Atlantic Ocean (D.P. indicates the Drake Passage, the narrowest part of the ACC). The arrows indicate water flow, and dashed lines signify water crossing constant-density surfaces, made possible by mixing. The upper shaded area is the warm water sphere, including the subtropical thermocline and mixed layer, and the lower shaded region is Antarctic Bottom Water. The bulk of the unshaded region in between is North Atlantic Deep Water.

years for a parcel to move from its high-latitude source to the equator, still longer if the path were not direct. Thus, if the surface conditions change, it will take several hundred years for the deep ocean to re-equilibrate.

OCEAN CIRCULATION IN A NUTSHELL

The large-scale ocean circulation may usefully be divided into a quasi-horizontal circulation, comprising the gyres and other surface and near-surface currents, and a meridional overturning circulation. Embedded within the circulation are smaller mesoscale eddies, which actually contain the bulk of the kinetic energy of the ocean and which are analogous to atmospheric weather systems.

The ocean gyres

- The ocean gyres are primarily wind driven, responding in particular to the north–south variations of the zonal wind. The subtropical gyres lie between about 15° and 45° in both hemispheres, with the subpolar gyres poleward of that in the Northern Hemisphere.
- The wind stress has a direct effect in the uppermost few tens of meters of the ocean, where it induces an Ekman flow at right angles to the wind. This Ekman flow in turn causes the sea surface to slope and produces a geostrophic flow, which is the main component of the gyres and which extends down several hundred meters.
- The main gyres all have a strong intense current at their western boundary (e.g., the Gulf Stream in the North Atlantic, the Kuroshio in the North Pacific), which arises from the combined effects of Earth's sphericity and its rotation.

The overturning circulation

- The overturning circulation is a response to variations in surface buoyancy, in that the densest water at the surface (usually at high latitudes) sinks and moves away from the sinking region at depth.
- For the circulation to persist, the deep water must be brought up to the surface; otherwise, the abyss will fill up with the densest water available and then stagnate. Two processes bring deep water up to the surface: mixing and the wind.

(continued)

— Mixing warms the deep water at low latitudes, which may then rise through the thermocline, maintaining a circulation of sinking at high latitudes and rising at low latitudes.

— Strong westerly winds in the Antarctic Circumpolar Current can draw water up from the deep and induce an interhemispheric circulation, which is particularly strong in the Atlantic.

The other main currents

• The Antarctic Circumpolar Current is the collection of eastward flowing currents around Antarctica, which taken together form the largest sustained current system on the planet. It is a response both to wind and to the meridional temperature gradient.

• The equatorial current systems are predominantly controlled by the winds, consisting typically of a westward flowing current and eastward countercurrents and undercurrents.

APPENDIX A: MATHEMATICS OF INTERIOR FLOW IN GYRES

Suppose that the wind blows zonally across the ocean, with a stronger eastward wind to the north, as in figure 4.3. Away from coastal regions (where friction may be important) the forces present are the zonal wind force (which here we simply denote F_w^x), the Coriolis force (fv and fu) and the pressure gradient force ($\partial\phi/\partial x$ and $\partial\phi/\partial y$, where $\phi = p/\rho$), and we represent their balance mathematically as

$$-fv = -\frac{\partial \phi}{\partial x} + F_w^x, \qquad fu = -\frac{\partial \phi}{\partial y}, \qquad (4.5a, b)$$

in the zonal and meridional directions, respectively. If there were no wind, the flow would be in geostrophic balance, and indeed the flow is in geostrophic balance at depths greater than 100 m or so, below the level at which the winds' effects are directly felt. Conservation of mass also gives a relation between u and v, namely

$$\frac{\partial u}{\partial x} + \frac{\partial v}{\partial y} = 0. \qquad (4.6)$$

If we cross-differentiate equation 4.5 (i.e., differentiate equation 4.5a with respect to x and equation 4.5b with respect to y and subtract), then the divergence terms vanish using equation 4.6 and the pressure gradient terms cancel, and we obtain

$$\beta v = -\frac{\partial F_w^x}{\partial y}, \qquad (4.7)$$

where $\beta = \partial f/\partial y$ is the rate at which the Coriolis parameter increases northward. The balance between the varying wind and the meridional flow embodied in equation 4.7 is known as Sverdrup balance, and the effect of differential rotation is called the beta effect. If the wind stress has a positive curl, that is, if $\partial F_w^x/\partial y > 0$, then, because β is also positive, v must be negative and the interior flow must be equatorward. There must be a poleward return flow in a boundary current at either the western or the eastern edge of the ocean basin, where the effects of friction conceivably can be such as to balance the Coriolis and wind stress curl terms. But only if the flow returns in the western boundary current can the frictional effects balance the wind stress curl overall, for then the flow overall has the same sense as the wind.

5 THE OCEAN'S OVERALL ROLE IN CLIMATE

....

The coldest winter I ever spent was
a summer in San Francisco.

—Mark Twain

THE OCEAN PLAYS A NUMBER OF ROLES IN OUR PRESENT climate, and in this chapter we discuss two of the most important:

1. The ocean moderates the climate by taking in heat when the overlying atmosphere is hot, storing that energy and releasing heat when the atmosphere is cold.
2. The ocean redistributes heat in the large-scale ocean circulation.

In addition, the ocean generally has a lower albedo than land, so that if all the ocean were replaced by land, the planet as a whole would be cooler. In some contrast, when the ocean freezes it forms sea ice, which has a generally high albedo. Thus, if the climate as a whole were to warm up, then the sea-ice extent would likely diminish, lowering the overall albedo and so further warming the planet. And finally, of course, the ocean is far and away the main reservoir of water on the planet, and if the planet were dry the atmosphere would have no clouds and the greenhouse effect would be much smaller, with wholesale changes in the climate. These last few effects are a little indirect, so let's focus on the two effects we mentioned first.

THE MODERATING INFLUENCE
OF THE OCEAN

Perhaps the most obvious effect that the ocean has on climate is its moderating effect on extremes of temperature, both diurnally (i.e., the day–night contrast) and annually (the seasonal cycle). We focus on the effects on the annual cycle because these tend to be on a larger scale and more befitting a book with *climate* in the title, but much the same principles and effects apply to the diurnal cycle. First we take a look at the observations to confirm that there *is* a moderating influence from the ocean. Fig. 5.1 shows the annual cycle of temperatures of San Francisco and New York. The two cities have similar latitudes (San Francisco is at about 38° N and New York is at about 41° N) and both are on the coast, yet we see from the figure that the range is enormously larger in New York—the highs are higher and the lows are lower. (One wonders if the respective climate extremes affect or even effect the different personalities of New Yorkers and Californians.)

The difference is mainly caused by the fact that the climate of San Francisco is *maritime,* meaning that it

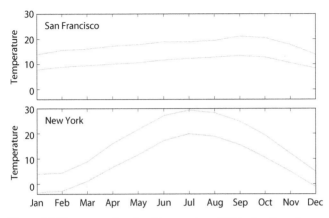

Figure 5.1. The seasonal cycle of temperature (°C) in San Francisco and New York. For each city, we plot the average low temperature and the average high temperature for each month. Note the much bigger range in New York and the maximum earlier in the year, in July rather than September.

is influenced by the ocean, whereas the climate of New York is, in spite of it being on the Atlantic coast, essentially continental. A city that is truly land-locked, such as Moscow, has a climate much more like New York than San Francisco. New York's climate is continental because the mean winds come primarily from the west, so they blow over land and take up its temperatures before they reach the city. In contrast, the winds have blown over the Pacific Ocean before arriving at San Francisco. So why does the ocean moderate the climate? It is in part because water has a relatively high heat capacity, compared to the material that makes land (e.g., soil and concrete), and in larger part because the upper ocean is in constant motion, and so the depth of ocean being heated and cooled over the seasonal cycle is much larger than the depth of land that is being heated and cooled. Let's explain that in a bit more detail.

First of all, the higher the heat capacity of a body, the more heat is needed to change its temperature. Thus, if an object is being cyclically heated and cooled, as in a seasonal cycle, then the change in its temperature is much smaller if its heat capacity is higher. Now, to what depth in the land and ocean does the heat penetrate over the course of a seasonal cycle? Plainly not all of the ocean's great depth is heated during the day or cooled during the night, or even over the course of a season. Rather, regarding the ocean, just that part that is turbulently mixed by the effects of wind (and in part by heating and cooling itself) fully partakes in the annual cycle, namely the *mixed layer,* which we discussed in chapter 2. Although its character varies from place to place in the ocean, it has a typical depth of about 50–100 m. That is, the effective heat capacity of the ocean is approximately that of a body of water 50–100 m deep. This heat capacity is quite large, and for comparison, the heat capacity of the atmosphere corresponds to a depth of just 3 m of water.

What is the effective heat capacity of land? Two effects make it much less than that of water. First, the specific heat of dry land is about 4 times less than that of water (for wet land, the factor is about 2). Second, because land is, rather obviously, not in motion in the same way as the ocean is, the penetration of heat into the land is much less than it is into the ocean. The heat can penetrate only by conduction, and because the earth (soil) has rather low thermal conductivity, only the top few meters are significantly heated and cooled over the course of a season. The same can be said for the major ice sheets over land, such as those over Greenland and Antarctica: they have large mass but low thermal conductivity. Thus, combining the effects of a larger heat capacity and a larger effective depth, the ocean has an effective heat capacity that is about 100 times greater than that of land. This high heat capacity considerably attenuates the seasonal cycle and is a good part of the reason for the large difference between San Francisco and New York.

San Francisco is a rather extreme case because not only is the summer temperature moderated by the ocean, but also the interminable fog that blows in from the ocean and covers the city like a wet blanket keeps the summer temperatures miserably low and makes them seem even lower, as Mark Twain perhaps felt. However, the heat capacity effect does occur on very large scales. The surface of the Southern Hemisphere is about 80% ocean whereas the Northern Hemisphere is only about 60% ocean, and as a consequence the seasonal cycle is much more pronounced over the Northern Hemisphere than the Southern, as we see in figure 5.2. Between 40° N and 60° N, the amplitude of the annual cycle is about 12°C, whereas between the corresponding latitudes in the Southern Hemisphere, the annual cycle has an amplitude of only 3°C.

....

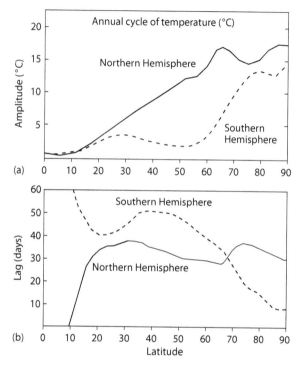

Figure 5.2. Amplitude and lag of the annual cycle in the Northern and Southern hemispheres, as a function of latitude. The lag is the time, in days, from the maximum solar insolation to the maximum temperature. Source: Trenberth, 1983.

The lag in the seasons

The observant reader noted in figure 5.1 that not only is the seasonal cycle more muted in San Francisco, but also that the maximum temperatures occur later in the season, in September. This again is an effect of the large heat capacity of the system, as a simple argument shows. Suppose that a system is heated externally (e.g., by the sun) and is cooled by the effects of longwave radiation and that the cooling is proportional to the temperature itself. If the system has a very small heat capacity, then the heating and cooling must balance each other at all times. A consequence of this is that the cooling is greatest when the heating is greatest, and so the temperature itself is highest when the sun is highest in the sky. Indeed, we find that in continental climates the temperature is highest fairly soon after the summer solstice and coldest soon after the winter solstice: In Fig. 5.1, we see that New York is hottest in July and coldest in January.

If a system has a large heat capacity, it takes some time to warm up and cool down, and so the maximum temperatures occur some time after the maximum insolation and thus later in the summer. The same effect occurs on a daily basis: inland, the maximum daily temperature occurs shortly after noon, whereas at the seaside the maximum temperature is later in the afternoon. On a large scale, in the Northern Hemisphere midlatitudes, the maximum temperature occurs on average about 30 days after the maximum solar insolation, whereas in the more maritime Southern Hemisphere, the maximum occurs about 45 days after peak insolation (figure 5.2). At very high latitudes, where the Southern Hemisphere is covered by land (Antarctica) but the Northern Hemisphere by ocean (the Arctic Ocean), the lag is longer in the Northern Hemisphere. A mathematical demonstration of this effect is given in appendix A of this chapter.

The general damping of climate variability by the ocean

Not only does the ocean provide a moderating influence on the march of the seasons, but it also can provide a moderating influence on the variability of climate on other timescales too. We talk more about the mechanisms that give rise to climate variability in the next chapter, but for now let us just suppose that the climate system excluding the ocean is able to vary on multiple timescales, from days to years. Then, just as the ocean is able to damp the seasonal variability, the ocean damps variability on all these timescales. However, the ocean does not damp the variations equally on all timescales; rather, because on longer timescales the ocean itself can heat up or cool down in response to climate variations, the damping effects are larger on shorter timescales. We give a brief mathematical treatment of this argument in the next section, and a more complete treatment in appendix A of this chapter.

Mathematical treatment of damping

The surface temperature of the ocean and the land are maintained by a balance between heating and cooling. The heating occurs both by solar radiation and by downward longwave radiation from the atmosphere and is proximately independent of the temperature of the surface itself. The cooling, on the other hand, increases with the temperature—a hot object cools down faster than a warm one. If for simplicity we suppose that the cooling rate varies linearly with temperature, then we can model the surface temperature by the equation

$$C\frac{\mathrm{d}T}{\mathrm{d}t} = S - \lambda T. \tag{5.1}$$

Here, S is the heating source, T is the temperature, and t, the time. The parameter C is the heat capacity of the system, and λ is a constant that determines how fast the body cools when it is hot. Obviously, this equation is too simple to realistically describe how the surface temperature varies (it ignores lateral variations, for one thing), but it illustrates the point we wish to make.

The equation says that the heat capacity times the rate of the temperature increase (the left-hand side) is equal to the heating (S) minus the cooling (λT). If we set $S = 0$ for the moment, then a solution of this equation is $T = T_0 \exp(-\lambda t/C)$, where T_0 is the initial temperature. If S is a constant, then the full solution is

$$T = \frac{S}{\lambda} + T_0 \exp(-\lambda t/C). \tag{5.2}$$

This equation tells us that if there is a perturbation to the system, the perturbation will decay away on the timescale C/λ. With a mixed-layer depth of 100 m and $\lambda = 15\ \mathrm{Wm}^{-2}\,K^{-1}$ (which is suggested by observations for air–sea interactions), we obtain a timescale of a little less than a year. That is to say, the ocean mixed layer can absorb or give out heat on the timescale of about a year. Variability on timescales significantly longer than this is not greatly damped by the presence of an ocean mixed layer because on these timescales the mixed layer itself heats up and cools down and so provides no damping to the system. However, on timescales much shorter than this, the mixed layer absorbs heat from a warm atmosphere, or alternatively gives up heat to a cold atmosphere, thus damping the variability that the atmosphere otherwise might have. The land surface has a much smaller heat capacity, so that the timescale C/λ is much smaller for land than it is for the ocean. There is thus a much smaller damping effect over land than over the ocean.

The situation is not *quite* as straightforward as this argument suggests. A complicating factor is that the entirety of the ocean mixed layer does not respond to fast variations in the atmosphere. Thus, for example, only the top few meters of water may respond to diurnal variations in temperature, and such variations are therefore damped less than one might expect. Nevertheless, the overall effect is clear: The heat capacity of the ocean mixed layer damps variations on timescales up to and including the annual variations. Interested readers can find a more complete description of this effect in appendix A of this chapter.

OCEAN HEAT TRANSPORT

The other great effect that the oceans have on the mean climate is that they transport heat, usually poleward, thus cooling the tropics and subtropics and warming high latitudes. Let's first look at how much the transport is, then we'll discuss the ocean processes that give rise to the transport, and finally what effects the transport has on climate.

How much?

On average, both the atmosphere and the ocean transport heat poleward, and this transport is illustrated in figure 5.3. The total transport of the atmosphere plus the ocean may be determined fairly directly from satellite measurements. Over the whole planet, there is a balance between the incoming solar radiation and outgoing longwave radiation, and if there were no heat transport, the incoming solar radiation would equal the outgoing infrared radiation at each latitude—a state of pure radiative balance. In fact, at low latitudes there is an excess of incoming solar radiation, whereas at high latitudes there is an excess of outgoing infrared radiation, meaning that at low (high) latitudes Earth is colder (warmer) than it would be if it were in pure radiative balance. The imbalance arises because heat is transported poleward by the motion of the atmosphere and ocean, and if we measure the imbalance at each latitude, then we obtain the total heat transport by the atmosphere and ocean. Perhaps needless to say, this measurement is easier said than done, but the advent of modern satellites that make separate measurements of solar and infrared radiation makes it possible. The most accurate estimates come from the period of the Earth Radiation Budget Experiment, in particular over the period 1985–1989, when intense observations were made, but data continue to be gathered.

The transport in Earth's atmosphere may be calculated directly because we are constantly taking measurements of the air temperature and its flow for weather forecasts. One way to then determine the total heat transport by the atmosphere at a given latitude is to sum up the product of the temperature and meridional velocity over all longitudes and over the entire depth of the atmosphere. Given the heat transport by both the atmosphere and by the atmosphere–ocean system, the heat transport by the ocean follows by simple subtraction, and this transport is shown in the dashed line in figure 5.3a. It is also possible to calculate the ocean transport directly, using in situ ocean measurements; the advantage is that one may be able to elucidate the individual mechanisms of ocean heat transport rather than just the overall effect. Such direct measurements tend to be less accurate than the residual method because of the sparsity of measurements in the ocean, but the two methods are broadly consistent.

Let's first look at the total heat transport. Evidently in mid- and high latitudes the atmospheric transport is two to three times that of the ocean, whereas in low latitudes the two are comparable, with the ocean exceeding that of the atmosphere at very low latitudes. The atmospheric heat transport, which we won't consider in any detail, takes place via two main mechanisms: In low latitudes, the transport occurs via the zonally symmetric Hadley cell, which takes warm air poleward and cooler air equatorward. In midlatitudes, the heat transport in the atmosphere occurs through the familiar weather systems,

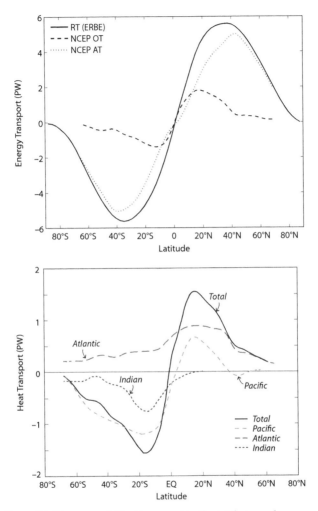

Figure 5.3. Upper panel: Heat transport in the total atmosphere–ocean system (solid line), in the ocean (dashed line), and in the atmosphere (dotted line). Lower panel: Oceanic heat transport, subdivided into the various basins. Source: Trenberth and Caron, 2004.

which are continually stirring the atmosphere and bringing warm air with low-latitude origins poleward and cool, high-latitude air equatorward.

As for the ocean overall, and as we might expect, it transports heat poleward in both hemispheres. The transport is associated with a release of heat into the atmosphere at high latitudes, whereas the ocean is being heated by the atmosphere at low latitudes. Measurements show that the release of heat occurs in two primary locations: at the poleward end of western boundary currents, notably in the western Atlantic and western Pacific oceans at about 40° north and south, and at very high latitudes, in particular in the North Atlantic around Greenland. Another notable feature about the ocean transport is that the poleward transport is much larger in the Northern Hemisphere than in the Southern. In fact, if we look at the contributions from the individual basins in the right-hand panel of figure 5.3, we see that the heat transport is northward (that is, *toward* the equator) in the South Atlantic! Such a transport is quite remarkable, for it implies that the ocean is not being thermally driven by the meridional temperature gradient alone (which would transport heat from hot places to cold places). What could be the driving force? If we recall our discussion of the oceanic meridional circulation in the previous chapter, we won't be too surprised to discover that it is the wind, but not the wind-driven gyres. Let's look at the mechanisms of ocean heat transport in a bit more detail and try to sort this out.

What mechanisms?

Oceanic heat transport is mainly effected by the large-scale circulation, with some transport by mesoscale eddies, mainly in the ACC. As we discussed in chapter 4, there are two distinct aspects to this circulation: the wind-driven gyres and upper ocean circulation, and the meridional overturning circulation. Let's see how each of these transport heat.

The wind-driven gyres

The wind-driven gyres, especially the subtropical wind-driven gyres, are a major factor in the heat transport in both hemispheres. If we consider the North Atlantic as an example, poleward heat transport occurs because the western boundary current (the Gulf Stream in this case) brings warm water up from the tropics along the eastern seaboard of the United States, releasing heat (especially in winter) when the warm water comes into contact with the colder air coming off the cold continental land mass. Such a release of heat occurs at the western edges of all the major ocean basins in midlatitudes, for example, off the coasts of Japan, the Eastern United States, South America (south Brazil, Uruguay, and Argentina) and southeastern Australia. The poleward flow of the western boundary currents is balanced by equatorial flow in the middle of the gyres that brings cold water equatorward, although the flow is broader and weaker so that the transport of cool water equatorward is spread over a large area. But, in any case, the consequence of warm water flowing poleward in the western boundary currents and cool water flowing equatorward in the interior means that the wind-driven gyres transport heat poleward. The transport occurs in both the Pacific and the Atlantic, and in both the Northern and Southern hemispheres. The subpolar gyres in the Northern Hemisphere also transport poleward, but they are less well defined than the subtropical gyres and cover less of the ocean so that their heat transport is somewhat weaker than that of the subtropical gyres.

The meridional overturning circulation

We saw in chapter 4 that the meridional overturning circulation has two mechanistically distinct components: a mixing-maintained component and a wind-maintained component, and the net overturning circulation is a combination of the two. The mixing-maintained component of overturning circulation responds to the buoyancy gradient at the surface between the equator and the pole, and in today's climate that buoyancy gradient is mainly a consequence of the temperature gradient. As described in chapter 4, the buoyancy gradient leads to a circulation in which cold water at high latitudes sinks and moves equatorward, balanced by warmer, near-surface water moving poleward. The net effect of this circulation is a poleward transport of heat that would be, in the absence of other effects, roughly equal in magnitude in the two hemispheres.

The other component of the overturning circulation is a pole-to-pole circulation, driven by the wind in the southern oceans. In the Atlantic Ocean, this component tends to dominate the purely buoyancy-driven circulation, as suggested by figure 4.8 in chapter 4. We see from this figure that there is a large northward, and so poleward, heat transport in the Northern Hemisphere because the equatorward moving water has come from high northern latitudes and is correspondingly cold, and the poleward moving water nearer the surface is warmer. In the Southern Hemisphere, however, the southward moving water is colder than the northward moving water because the cold North Atlantic Deep Water continues its path into the Southern Ocean and the water moving northward near the surface is relatively warm. That is, the water moving equatorward is generally warmer than the water moving poleward! Thus, the heat transport in the Atlantic from the deep circulation is northward in both hemispheres. The wind-driven gyres, of course, transport heat poleward in both hemispheres, partly counteracting the equatorward heat transport by the deep circulation in the Southern Hemisphere, so that the net effect is that the heat transport is weak and equatorward in the South Atlantic, whereas it is strong and poleward in the North Atlantic. In the Pacific and Indian ocean basins, there is no corresponding wind-driven deep circulation that transports heat northward, and the wind-driven and buoyancy-driven circulations both act to transport heat poleward, as we can see in figure 5.3.

What are the effects?

What are the gross effects of the ocean heat transport on the climate? The main effect is simply that the high latitudes, especially the high latitudes in the Northern Hemisphere, are warmer than they would be if the oceans were not present. How much warmer is a question that we cannot answer with armchair reasoning. We would need to perform detailed calculations with comprehensive climate models of the type used to predict the weather or used for global warming experiments.

One such set of experiments was performed by M. Winton of the Geophysical Fluid Dynamics Laboratory in Princeton, and we briefly describe some of the results found (Winton 2003). Climate models solve the equations that determine the temperature and motion of both the atmosphere and the ocean. The models also have representations of sea ice and cloudiness and of their effects on the incoming solar radiation and outgoing infrared radiation. Thus, for example, snow and ice cover reflects solar radiation back to space, making the climate cooler than it would be in their absence. If the ice sheets were for some reason to expand, the climate would cool, the ice sheets would further expand, and the climate would further cool—an example of a positive feedback, in this case the *ice–albedo feedback*.

In one set of numerical experiments, the ocean was replaced by a mixed layer, so that although the heat uptake and release of the ocean throughout the annual cycle are accounted for, the effect of the ocean currents and so the oceanic heat transport are completely removed. When the ocean heat transport is removed, the atmosphere tries to compensate for this change by transporting more heat poleward itself. However, in spite of this and in spite of the relatively small heat transport of the oceans compared to the atmosphere at high latitudes, the effects of the oceans are found to be quite large because of a feedback involving sea ice and, to a lesser extent, low-level cloudiness. The simulations without oceanic heat transport all developed large ice sheets that covered mid- and high latitudes, making the overall climate much colder than it is now.

What seems to happen is the following. Although the atmosphere is able to partially compensate for the lack of an ocean transport, the atmospheric transport naturally occurs at a higher elevation than the ocean transport. The lack of an ocean heat transport enables sea ice to grow, and once the ice begins to grow, the positive ice–albedo feedback comes into play and the ice grows more. The detailed mechanism for the strong effect of the ocean seems to involve the upward convective flux of heat in the wintertime: The meridional overturning circulation leads to convection at high latitudes, with cold water parcels sinking and being replaced by slightly warmer parcels, which then release heat into the atmosphere. This process is eliminated when the ocean is replaced by a mixed layer, allowing sea ice to grow.

The Gulf Stream and the climate of Britain and Ireland

It is occasionally said that Britain and Ireland owe their mild climate to the presence of the Gulf Stream and the North Atlantic Drift. Thus, the story goes, the Gulf Stream brings warm water from Florida up the eastern seaboard of the United States and then across the Atlantic in the North Atlantic Drift to the shores of Britain and Ireland, hence moderating the otherwise cold winters. Certainly, the surface temperature of the eastern North Atlantic is a few degrees warmer than the water at the same latitude off the coast of Newfoundland, as figure 2.2 in chapter 2 illustrates. Although this difference does have some effect on the temperature differences between the two locations, Britain and Ireland have a moderate winter climate primarily as a consequence of the fact that they are next to the ocean, with the ocean on their west. Even if there were no gyres in the ocean at all, the climate of these parts would be much more moderate than the climate at similar latitudes on the eastern sides of continental land masses. Thus, Britain and Ireland have a much more similar climate to British Columbia, at a similar latitude on the west coast of Canada, than they do to Newfoundland and Labrador on the east coast. The effects of the east–west asymmetry of sea-surface temperatures on the seasonal climate of Britain and Ireland, and of midlatitude coastal areas surrounding the ocean basins generally, are relatively small.[1]

However, the effects of the ocean and the ocean circulation on the climate of Britain and Ireland are far from small. If the ocean were to cease circulating altogether—that is, both the gyres and the meridional overturning circulation were to cease—then the high latitudes would generally get colder, as we discussed in the previous section, and possibly freeze over. If the oceans did not freeze, western Europe would still have a maritime climate and a more moderate seasonal cycle than the eastern United States and eastern Canada.

APPENDIX A: THE MATHEMATICS OF THE RELATIONSHIP BETWEEN HEATING AND TEMPERATURE

In this appendix, we give an elementary mathematical treatment of the relationship between heating and temperature. We will explain two things: why the temperature range is smaller if a body has a larger heat capacity and why there is a lag between heating and temperatures.

We model the system with the simple equation

$$C\frac{dT}{dt} = S - \lambda T. \tag{5.3}$$

Here, S is the heating source, T is the temperature, and t, the time. The parameter C is the heat capacity of the system, and λ is a constant that determines how fast the body cools when it is hot. The equation says that the heat capacity times the rate of the temperature increase (the left-hand side) is equal to the heating (S) minus the cooling (λT). Let us further suppose that the heating is cyclic, with $S = S_0 \cos \omega t$, where ω is the frequency of the heating and S_0 is its amplitude.

To solve the equation, we write $S = \text{Re } S_0 \exp(i\omega t)$ and seek solutions of the form $T = \text{Re } T_0 \exp(i\omega t)$ where Re means "take the real part" and T_0 is a constant to be determined. Substituting into equation 5.3, we have

$$CT_0 i\omega e^{i\omega t} = S_0 e^{i\omega t} - \lambda T_0 e^{i\omega t}, \tag{5.4}$$

where only the real part of the equation is relevant. From this equation, we straightforwardly obtain

$$\begin{aligned} T &= \text{Re } \frac{S_0(\lambda - iC\omega)e^{i\omega t}}{\lambda^2 + C^2\omega^2} \\ &= S_0 \frac{\lambda \cos \omega t + \omega C \sin \omega t}{\lambda^2 + C^2\omega^2}. \end{aligned} \tag{5.5}$$

What does this equation tell us? If the heat capacity is negligible, or if the frequency ω is very small (i.e., very slow variations in forcing), then

$$T \approx \frac{S_0}{\lambda} \cos \omega t. \tag{5.6}$$

The temperature is in phase with the heating, and the amplitude of the cycle is S_0/λ. If the heat capacity is large or the frequency is high, then

$$T \approx \frac{S_0}{C\omega} \sin \omega t. \tag{5.7}$$

The amplitude of the cycle is $S_0/C\omega$, which is small because C is large. That is, because $C\omega > \lambda$ in this case, the variations in temperature are smaller than they are in the case with slow forcing variations given by equation 5.6. Because high frequencies are damped more than low frequencies, we say that the variations are *reddened*. Note too that the temperature variations are now out of phase with the heating. Thus, the maximum temperature occurs when the heating is least. It is this effect that accounts for the delay in the maximum temperature in maritime climates, with the hottest part of the year occurring in late summer or even the beginning of autumn.

6 CLIMATE VARIABILITY FROM WEEKS TO YEARS

···

Climate is in the eye of the beholder.

IN THIS CHAPTER WE LOOK AT CLIMATE VARIABILITY,
and in particular climate variability that is associated in
one way or another with the ocean. This condition is not
very restrictive because nearly all forms of climate variability on timescales of months to decades are affected
by, or even caused by, the ocean. Even in cases in which
the underlying cause of the variability is nonoceanic, the
ocean may modulate the variability and determine its timescale, and in many ways we can think of the ocean as the
pacemaker of climate. We won't talk about climate variability on timescales of centuries to millennia and longer,
for that deserves a book on its own. Rather, our focus will
be on midlatitude climate variability on intraseasonal and
interannual timescales, and on climate variability associated with El Niño. Before delving into all that, let's first
discuss if and how climate is different from weather.

CLIMATE AND WEATHER

What is the difference between climate and weather? It is
intuitively clear what the weather is—it is the day-to-day

state of the atmosphere at some location, usually with particular reference to such things as temperature, windiness, and precipitation. It is also intuitively clear that when we speak of climate we wish to average out all these day-to-day fluctuations and refer to some kind of average of the weather. But what precisely? There is no ideal definition of climate, but a useful working notion is that climate is the statistics of the weather—the mean, the standard deviation, and so forth. However, this notion slightly begs the question of how we calculate the statistics—how do we take the mean, for example? And if climate is a time average of the weather, then how can climate have any temporal variability?

Although it is rather fanciful, it is useful to envision a thought experiment in which we take the *ensemble average* of the weather. Thus, we envision a large number of identical planet Earths, forced the same way, but each one started out in a slightly different way so that each has different weather. We could then unambiguously define the climate to be the average, along with other relevant statistical quantities like the variance, over the ensemble of planet Earths. If the forcing were to change, perhaps because the CO_2 levels in the planets were to increase, then the climate of the ensemble would also change.

The problem with this definition is that it is not practical, there is no such ensemble in reality. However, we can try to take our average in such a way that it mimics the ensemble average as closely as possible, and this way of proceeding will be useful to the extent that the weather and climate have different time- and space scales. We could then define climate as the average of the weather over a time period long enough so that weather fluctuations are averaged out but variability on longer timescales is still allowed. Weather typically varies on timescales of a few days to a few weeks, so that we might define climate as the average weather over time periods longer than a month, say. In practice, this time period is too short for many purposes because the monthly average temperature still fluctuates considerably, and a more common definition takes the climate to be the average (along with other statistical quantities) over a period of a few years, with the precise averaging period depending on what quantity is of interest. We might choose to take the average over a particular time of year—only over the winter months, for example, and so obtain a winter climatology. If we are interested in how climate varies across ice ages, then averaging over a period of centuries or even millennia might be appropriate, but if we are interested in whether climate changed over the course of the twentieth century, a much shorter period is obviously more appropriate.

The moral of the above discussion is that, although it is useful to think of the climate as some kind of average, there is no compact single definition of climate that is useful and appropriate for all purposes, and we are often better served by talking about the climate with reference to a particular timescale. Climate varies on more than one timescale—indeed, there may be no timescale on which we can say there is no climate variation.

CLIMATE VARIABILITY IN MIDLATITUDES

Weekly variability and the weather

On daily and weekly timescales, the main mechanism that causes variations in the temperature and wind is simply the familiar *weather*. The mechanism that gives rise to weather resides in the atmosphere, and it is the consequence of a fluid instability called *baroclinic instability*. This instability can be thought of as a type of convective instability in which if a fluid is heated from below it expands, becomes lighter than its surroundings, and therefore rises. Baroclinic instability has a similar origin, but it also involves the lateral motion of fluid parcels, with a typical scale of up to a few thousand kilometers and a typical velocity of about 10 m s^{-1}. It takes a parcel moving at this speed a little more than a day to travel 1,000 km, hence accounting for the typical timescale of days to weeks. In midlatitudes, the average temperature difference between two lines of latitude that are 1,000 km apart is about 5°C, and because weather is stirring the air on these spacescales, a typical weather system can lead to typical temperature variations of up to 5°C on timescales of a few days.

Monthly to seasonal variability

As we discussed in chapter 1, the seasonal cycle itself arises because of the tilt of Earth's axis of rotation relative to the Earth–sun axis. The axis of rotation of Earth is fixed relative to the distant stars, so that as Earth moves around the sun, the North Pole points generally toward the sun (giving the Northern Hemisphere summer), away from the sun (giving the Northern Hemisphere winter) or somewhere in between, as illustrated in figure 1.2 in chapter 1. If we think of a season as lasting roughly three months, are there any climate phenomena that have timescales between the weather timescale and the seasonal timescale? The answer is, to a degree yes, there is at least one such phenomenon, known as the *North Atlantic Oscillation* (NAO). The NAO is a phenomenon at the interface between weather and climate that dictates variability on a monthly timescale over the North Atlantic and surrounding regions, thus from the eastern seaboard of the United States, over Greenland, to Europe, and from the Arctic region to the Canaries. Indeed, to some extent the NAO affects the weather and climate over the entire Northern Hemisphere. There is an analogue of the NAO in the Southern Hemisphere (called the Southern Annular Mode), although it has a more hemispheric extent, and a somewhat similar pattern in the Pacific Basin (called the Pacific–North American pattern), but none are quite as well studied as the NAO, so let us focus on that.

So what is the NAO? It may be thought of as a north–south oscillation of the main patterns of weather variability over the Atlantic region, especially in the winter, as illustrated in figure 6.1. During the positive phase of the NAO, the main path of weather systems tracks a little further north than usual. Because the air is coming

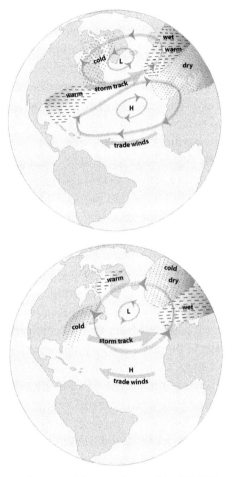

Figure 6.1. A schematic of the two phases of the NAO. The top panel shows the positive phase of the NAO, with storms tracking northward bringing mild but wet weather to northern Europe. The bottom panel shows the negative phase, with storms tracking southward bringing wet weather to southern Europe and colder weather to northern Europe.

straight over the ocean, it tends to be relatively warm and so brings warm, wet weather to the United Kingdom and other parts of northern Europe, with precipitation often falling as rain rather than snow in the United Kingdom, with similar effects downstream into eastern Europe and even Asia. Southern Europe and North Africa tend to have somewhat cooler weather than usual during these periods. Meanwhile, stronger northerly and northeasterly winds over Greenland and northeastern Canada bring cold, dry air to these parts, decreasing the temperature, with the eastern parts of the United States getting higher temperatures and more precipitation than normal, rather like northern Europe.

During the negative phase of the NAO, the storm track swings southward, bringing mild, wet weather to southern Europe. During these periods, northern Europe tends to receive air that has come from the east, which, since it has come from a continental land mass in winter, tends to be very cold, with precipitation often in the form of snow. Greenland, on the other hand, is now somewhat warmer than usual. The signal of the NAO is evidently quite large and coherent and in fact accounts for about one-third of the Northern Hemisphere's interannual surface variance during winter.

The distinctive pattern of the NAO, whether positive or negative, tends to last for several days to a few weeks, perhaps a little longer than regular weather patterns, but undoubtedly the main mechanism for the NAO lies in the atmosphere itself. However, there may be a longer timescale to the NAO, as illustrated in figure 6.2. Large

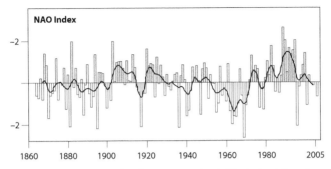

Figure 6.2. The NAO index from 1860 to 2005. The index is based on the normalized difference in average winter surface pressure between Lisbon, Portugal, and Stykkishólmur, Iceland. The heavy solid line shows the index after it has been smoothed to remove fluctuations of less than four years.

changes in the NAO index occur from year to year, and on the whole atmospheric behavior in one winter is largely independent of its behavior the previous winter. (The correlation in the NAO index from one year to another is only about 0.1.) Nevertheless, there is some indication that there are certain periods of time when the NAO persists from year to year. For example, from about 1905 to 1915 the NAO index was largely positive. It was negative from about 1950 to 1970, and then unusually positive in the 1980s and 1990s. Among other things, the positive index in the 1980s and 1990s brought higher than usual precipitation to Scandinavia and may have ameliorated the effects of global warming in the retreat of the glaciers. In contrast, over the Alps, a positive NAO index and less than normal precipitation may have worked *in conjunction* with global warming to cause a significant retreat of the Alpine glaciers. The Iberian peninsula and other areas around the Mediterranean also experienced drought in the late twentieth century.

It is an open question whether such long periods of persistent behavior are much more than the random variations of a chaotic system (rather like throwing six sixes in a row with a die) or whether they have a specific cause, most likely in the ocean. Why is the cause most likely in the ocean? It is because the timescales of the interannual variability correspond most closely to the timescales occurring in large-scale ocean dynamics, for example, in the dynamics of the ocean gyres. The dominant internal dynamics of the atmosphere are most likely too short to produce coherent dynamics that last over decadal timescales, whereas the dynamics of land ice sheets, or of changes in insolation at the top of the atmosphere because of orbital variations, are too long. One possibility is that volcanoes can have a climatic effect on the decadal timescale, but the correlation of the NAO with volcanoes is small. So let us look at the role of the ocean—that is, after all, the topic of this book.

Ocean influence on climate variability

How might the ocean affect climate variability? Indeed, might it even *effect* climate variability? The ocean has one unambiguous influence on midlatitude climate variability and a number of more ambiguous influences. The unambiguous influence stems from the fact that the heat capacity of the ocean is much greater than that of land, as we discussed in chapter 5. Thus, changes in surface temperatures over land, especially in locations far from the ocean, are much larger than the changes over the ocean, and so the variations over land tend to dominate any hemisphere-wide measure of the variability. Changes in the temperature associated with changes in the NAO pattern from positive to negative are distinctly more marked over land than over the ocean.

The more ambiguous effect concerns the relationship between fluctuations in the sea-surface temperature (SST) and the state of the atmosphere in midlatitudes, and in particular the state of the NAO. Certainly, variations in the SST and the overlying atmosphere *are* related. To be specific, let us focus on the North Atlantic and the NAO, but similar concerns and effects almost certainly apply to other regions of the world's ocean. It turns out that a common pattern of SST variability in the North Atlantic winter is a tripole. Rather like the NAO itself, the tripole commonly exists in one of two phases. In one phase, the pattern consists of a cold anomaly in the subpolar North Atlantic around Greenland, a warm anomaly in the middle latitudes off the Atlantic seaboard of the United States, and a cold subtropical anomaly between the equator and 30° N, concentrated most in the eastern Atlantic. Is the SST pattern created by the pattern of variability of winds and temperature in the atmosphere, or does the SST pattern determine the variability of the atmospheric winds and temperature? This is the $64,000 question! It may of course be a chicken-and-egg problem, with one pattern leading to the other, which then reinforces the first pattern, and so on. How are we to determine the answer? One way is to see if we can determine if variations in the atmosphere unambiguously lead those in the ocean, or vice versa. If it is the former case, then it seems likely that the atmosphere is driving the ocean, rather than vice versa (although of course the reader may be able to come up with perverse counter-examples where this is not the case).

Careful observational analysis in fact suggests that the basic pattern of SST anomalies is created by air–sea heat exchanges and the wind-induced near-surface ocean currents associated with the NAO. It seems that the correlations between atmosphere and ocean are strongest for atmospheric patterns that exist before the SST variability by a few weeks, suggesting that large-scale SST patterns are responding to atmospheric forcing, rather than causing the atmospheric patterns. Put simply, the atmosphere leads the ocean. However, rather intriguingly, that may not be the whole story, although the reader should be warned that the situation is far from settled and is an active topic of research. At still longer timescales, there is some evidence that a large-scale, pan-Atlantic SST pattern actually *precedes* the atmospheric NAO pattern by up to about six months.[1]

What is going on? We can explain these apparently differing observations as follows. First of all, let's be clear that we are talking about patterns of *variability* in both the atmosphere and the ocean. The mean ocean gyres and the meridional overturning circulation are set up, as we discussed in chapter 4, in response to the mean atmospheric circulation. Now, as we know, the atmosphere varies on timescales of days to a few weeks, and on space scales of a few thousand kilometers, and we call this variability weather. It seems that on timescales from days to a few months, the atmosphere does indeed drive anomalies in the ocean. That is to say, if the atmospheric winds and temperature happen to have a certain configuration for a few days, then (depending on that configuration) cold or warm SST anomalies can be generated, typically also on space scales of a few thousand kilometers. Because of the much larger thermal inertia of the ocean (as well as the fact that the ocean currents are roughly a hundred times smaller than the winds in the atmosphere), these patterns may persist much longer than typical atmospheric patterns. Thus, the ocean smooths away the smaller scale, day-to-day variability and leaves behind larger scale and more persistent patterns. The ocean is really a passive partner in this activity—the origins of patterns of SST variability on monthly timescales primarily lie in the atmosphere—and in this sense we may say the atmosphere drives the ocean. The persistence of the ocean anomalies does feed back on the atmosphere, leading to somewhat more persistent and less extreme weather than might otherwise be the case, so that one might also say the ocean damps the atmosphere. This effect is obviously rather similar to the one we discussed in the previous chapter, where we discussed the generally moderating influence of the ocean on the climate.

....

However, the ocean does have dynamics of its own. The gyres contain smaller eddies that produce considerable variability, just as the atmosphere has weather, and the large-scale gyres themselves may vary on interannual to decadal timescales. The meridional overturning circulation also varies, and its sluggish nature suggests that the variability may include timescales of decades and longer. If this variability is able to produce large-scale, long-lived SST anomalies, it is possible that these anomalies may, over time, affect the average behavior of the atmosphere. Even though the impact of mid- and high-latitude SST anomalies on the atmosphere seems to be small, if the anomalies persist for long enough, they will have a cumulative effect. Such an effect may be responsible for the seeming persistence of the NAO pattern in certain decades that we alluded to previously, but the evidence is not definitive.

EL NIÑO AND THE SOUTHERN OSCILLATION

In the next few sections we describe the phenomenon (and what a phenomenon it is!) known as *El Niño*, or sometimes as El Niño and the Southern Oscillation (ENSO). The Southern Oscillation is the atmospheric part of the phenomenon, and El Niño, the oceanic component; ENSO is the combination. However, unless we need to be particularly precise, we often just use the term El Niño, for this has a pleasant euphony lacking in the acronym ENSO.[2] El Niño is the largest and most important phenomenon in global climate variability on interannual timescales. In brief, it is an anomalous warming of the surface waters in the eastern equatorial Pacific, most pronounced just off the coast of Peru but extending westward to the dateline.

In the next few sections we describe in a little more detail what El Niño is, what mechanisms give rise to it, and what effects it has, or is perceived to have, on weather and climate throughout the world.

What is El Niño?

Every few years the temperature of the surface waters in the eastern tropical Pacific rises quite significantly. The strongest warming takes place between about 5° S and 5° N, and from the west coast of Peru (a longitude of about 80° W) almost to the dateline, at 180° W, as illustrated in figure 6.3. The warming is significant, with a difference in temperature up to 6°C from an El Niño year to a non–El Niño year. The warmings occur rather irregularly, but typically the interval between warmings ranges from three to seven years, as illustrated in figure 6.4.

The warmings have become known as El Niño events, or even (with a little violence to the Spanish language) El Niños. The name derives from the fact that the warm waters off Peru appear at about Christmastime, and the name el Niño is Spanish for the Christ child.[3] The warmings typically last for up to a year, sometimes two, and appear as an enhancement to the seasonal cycle, with high temperatures appearing at a time when the waters are already warming. Although there is no universally

(a) Normal (December 1996)

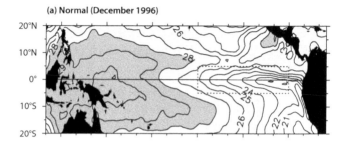

(b) El Nino (December 1997)

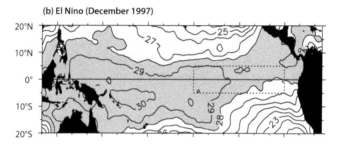

(b)–(a)

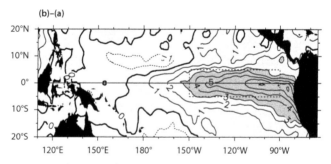

Figure 6.3. The sea-surface temperature in December of a normal (i.e., non–El Niño) year (December 1996, top panel); in a strong El Niño year (December 1997, middle panel); and their difference (bottom panel). A normal year is characterized by a cold tongue of water in the eastern tropical Pacific, which disappears in El Niño years.[5]

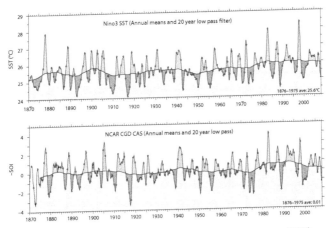

Figure 6.4 Top: A time series of the sea-surface temperature (SST) in the eastern equatorial Pacific region (specifically, in the so-called Niño 3 region). The spiky curve shows the annual means, and the dots represent Decembers. The smoother curve shows the SST after the application of a 20-year low-pass filter, and the top of the gray bar is the 1876–1975 mean. Bottom: A similar plot for the negative of the Southern Oscillation Index (SOI), the anomalous pressure difference between Tahiti and Darwin. Particularly large El Niño events can be seen in 1877–78, 1982–83, and 1997–98.[6]

agreed-upon definition of an El Niño, an event is often regarded as having occurred when there is a warming of at least 0.5°C averaged over the eastern tropical Pacific lasting for six months or more.[4] Rarely can a year be described as truly normal; rather, the ocean temperatures tend to fluctuate between warm El Niño years and years in which the equatorial ocean temperatures are colder in the east and warmer in the west, with those years that are particularly anomalous this way having become known as *La Niña* events (la niña, without capitalization, is Spanish for young girl; there is no female equivalent of the Christ child, el Niño in Spanish). We have direct observational evidence—that is, compilations of measurements of the sea-surface temperature from ships and buoys—of El Niño events for more than a century, but the events have almost certainly gone on for a much longer time, perhaps millennia or longer. We know this through a variety of proxy records; tree rings provide some of the most detailed information. As we will discuss later on, El Niño events bring anomalous rainfall and temperature throughout the equatorial Pacific and western North America, affecting the tree-ring characteristics and providing convincing evidence of the occurrence of El Niño events for the past several hundred years.

Corals also provide a good record of El Niño events because they have skeletal growth bands that, rather like tree rings, provide an accurate annual chronology. The isotopic composition of oxygen contained in the skeletons responds to both SST and rainfall, thus providing a record of El Niño that goes far into the past. Indeed, one recent analysis of corals from Papua New Guinea suggests that El Niño events have occurred for the past 130,000 years—that is, even over the last ice age![7]

Our knowledge of El Niño seems to have begun with observations of the SST off the coast of Peru, but it is now understood that the phenomenon itself is Pacific-wide and also involves the atmosphere. Observations of the winds and the surface pressure in the equatorial Pacific show that these tend to covary with the SST. In particular, during El Niño events the equatorial Pacific trade winds, which normally blow toward the west, become much weaker and may even reverse. One useful measure is the pressure difference between Darwin, at the far north of Australia (12° S, 130° E), and Tahiti (17° S, 150° W), an island in the Pacific; the record of this difference is known as the *Southern Oscillation.* (There is nothing truly special about these locations vis-à-vis El Niño, but pressure measurements have been made there for a long time.) We see in figure 6.4 that the Southern Oscillation and the SST record of El Niño are highly correlated.

The mechanism of El Niño

The mean state of the atmosphere and the ocean

Before discussing what processes conspire to produce El Niño events, let us discuss what the mean state of the atmosphere and ocean are, beginning with the atmosphere. The trade winds throughout the equatorial region blow predominantly from higher latitudes toward the equator, and from the east to the west. The low-level convergence at the equator forces the air to rise and then, several kilometers above the surface, move poleward, sinking in the subtropics at about 30° north and south and then moving equatorward at the surface. The meridional cell is known as the Hadley cell, and if there were no continents, the equatorial convergence would be on average the same at all longitudes.

However, there *are* continents and ocean basins, and their presence is essential in producing El Niño events. The easterly winds over the equatorial Pacific tend to push the surface waters westward, creating a westward current at the surface. The surface waters also diverge away from the equator because of the Coriolis effect, causing upwelling along the equator; that is, cold water from the deep ocean is pulled toward the surface to replace the surface waters moving away. Upwelling in fact occurs at most longitudes in the Pacific (because the equatorial divergence occurs at all longitudes), but it is particularly important in the east because here it must also replenish the surface waters moving westward away from South America. The upwelling water is cold because it comes from the deep ocean, so that the SST of the eastern equatorial Pacific is relatively low, and the surface waters become warmer as they move westward. The thermocline deepens further west, and so the upwelling does not bring as much cold abyssal water to the surface. The upshot of all this is that the SST is high in the western Pacific, up to about 30°C, and low in the eastern Pacific, at about 21°C (figure 6.3, top panel).

As one might expect, such a strong temperature gradient in turn affects the atmosphere. The warm western Pacific region becomes much more unstable with respect to convection than the eastern part, so that rising motion preferentially occurs here. The eastern Pacific is correspondingly cool and thus prone to descent and so, as illustrated in figure 6.5, an east–west overturning circulation is set up in the atmosphere. This is known as the

Walker circulation after the British meteorologist Gilbert Walker, who described it in the 1920s.[8] The Walker circulation coexists with and is somewhat analogous to the north–south Hadley circulation, in that both are driven by horizontal temperature gradients at the surface.

Variability and El Niño

Having described the Walker circulation and the corresponding ocean circulation, let's see if we can understand why they vary and give rise to El Niño events. We first note that the ocean and the atmosphere tend to reinforce each other: the atmospheric winds naturally blow from east to west, which sets up an east–west temperature gradient in the ocean. The warm ocean in the west leads to rising motion in the atmosphere, pulling air in at the surface. Thus, the westward atmospheric winds in the Pacific tend to be stronger than they would be if there were no ocean there at all or if the ocean extended all the way around the globe. Such a reinforcement is known as a *positive feedback,* and it can work both ways. That is to say, positive feedbacks tend to reinforce tendencies, so that if a system starts moving in a different direction, the feedback may again kick in, reinforcing the initial tendency, whatever that tendency may be.

Let's suppose that the Pacific is in a relatively normal state, as illustrated in the top panel of figure 6.5. We now imagine that, for reasons we won't be definitive about right now, the east–west temperature contrast weakens somewhat—perhaps the eastern Pacific Ocean gets a bit

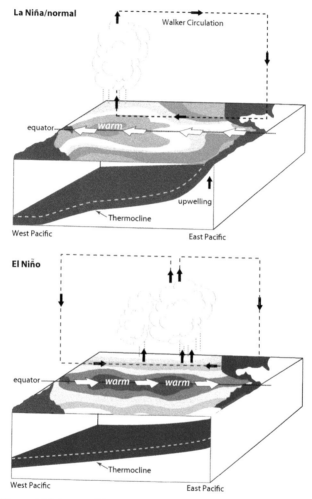

Figure 6.5 Schema of the atmosphere and ocean at the equator in the Pacific, during La Niña/normal conditions (top) and El Niño conditions (bottom). La Niña conditions are similar to normal conditions but with a still steeper-sloping thermocline and maximum SSTs a little further west.

warmer than usual because of some fluctuation in the ocean. The atmosphere can be expected to respond to this by a weakening of the trade winds, to which the ocean in turn responds; with weaker trade winds, the upwelling in the east becomes weaker and the thermocline flattens (as in the lower panel of figure 6.5). The temperature in the east consequently rises a little while temperature in the western Pacific falls, so further reducing the east–west temperature gradient. This reduced temperature contrast results in the main convective regions moving further east and causes the trade winds to further weaken, which in turn further reduces upwelling and causes the temperatures to further increase in the eastern equatorial Pacific; the culmination of this chain of events is an El Niño. Evidently, the sequence of events is crucially dependent on the mutual interaction and feedback between the atmosphere and the ocean, as was first posited by Jacob Bjerknes in 1969. (Jacob Bjerknes (1897–1975) was a Norwegian climate scientist working in the United States.)

The positive feedback we've described must come to an end eventually. The ocean temperature in the east cannot rise indefinitely (and in practice much above 30°C) because eventually the ocean cools via radiation and by giving up heat to the atmosphere. Furthermore, the natural tendency of the trade winds is to blow toward the west, and this effect tries to restore the natural order of things. Once an El Niño is fully developed, it persists for several months before beginning to decay. At that stage, the feedbacks we described above come into play again, but now working in the opposite direction. Indeed, the feedbacks may well cause the system to overshoot its mean state and go into a La Niña state, with enhanced warming in the west, cooler than normal conditions in the east, and stronger than usual trade winds. Then, some time after that, the whole process may begin again and the system may again evolve toward an El Niño state. The seemingly oscillatory nature of the atmospheric cycle historically gave rise to the appellation Southern Oscillation, and the entire phenomenon—atmosphere plus ocean together—is often referred to as the ENSO cycle. Although the basic mechanisms of the oscillation are fairly well accepted, the nature and causes of the transition between the El Niño and La Niña states, and the need or otherwise for some form of kick start to begin the El Niño cycle, remain to be definitely explained.[9]

Consequences and impacts of El Niño

What are the consequences of El Niño? There are both local effects, arising because of the warm eastern Pacific, weakened trade winds, and shifted regions of convection, and distant effects, mostly arising because the atmosphere carries the signal of El Niño far afield. We'll talk about the local effects first.

Local effects of El Niño

El Niño brings significantly warmer water to the eastern tropical Pacific, and this water spreads both north and south along the coast, giving a detectable signal in the SST as far north as California and Oregon. There is a direct effect of this change on Pacific marine life: Warm-water species have an extended range of habitat, whereas cold-water species try to move poleward or into deeper water, and the reduced upwelling off the South American coast produces fewer nutrients, fewer fish, and fewer sea-birds that feed off the fish. The shift of fish populations is a problem for the fishing industry, if only because the species tend to move away from established fisheries, and overall there is a loss of commercially important species. Pinnipeds (e.g., fur seals and sea lions) are also affected as far north as California (for example, in the Channel Islands off Los Angeles) because they may have difficulty finding adequate food supplies. Coral bleaching (the whitening of corals because of the loss of or changes in the protozoa living within them) in the Galapagos also occurs in El Niño years.

The warm ocean also affects the atmosphere above it; the warming is part of the ENSO cycle itself. During a warm event, the main convective areas move eastward, and the western parts of South America, especially close to the equator in northern Peru and Ecuador, experience significantly more rainfall, especially from December to February, when the events typically reach their peak. Significant flooding may occur when the event is a strong one. In contrast, northern Australia and the Indonesian archipelago experience significantly less rainfall than normal throughout the entire region and for an extended period of time. During a La Niña, the situation

EL NIÑO IN A NUTSHELL

1. El Niño refers to the warming of the surface waters in the eastern equatorial Pacific. The interval between warm events is typically from two to seven years but is quite irregular and can be longer.
2. El Niño events are associated with a weakening of the trade winds and a shifting eastward of the region of convection. The atmospheric side of the events is known as the Southern Oscillation, and the entire phenomenon is known as the El Niño–Southern Oscillation, or ENSO.
3. In contrast to an El Niño event, from time to time the far western equatorial Pacific becomes warmer than usual and the eastern Pacific, colder, which is known as a La Niña event.

What causes El Niño events?

1. El Niño is caused by the mutual interaction between the atmosphere and ocean in the equatorial Pacific, involving a positive feedback between the sea-surface temperatures and the strength of the trade winds.
2. The trade winds normally blow westward, but during an El Niño event the trade winds weaken, allowing the temperatures in the eastern equatorial Pacific to rise, further weakening the trade winds and so on.

What are some consequences of El Niño?

1. The global surface temperature in an El Niño year is up to 0.5°C higher than normal.
2. Convective rainfall in North Australia and Indonesia is suppressed in El Niño years, whereas rainfall is enhanced in western tropical South America.

3. During strong El Niño events, the atmospheric storm track in the eastern Pacific is stronger and further south than normal, bringing heavy rain to central and southern California and penetrating inland across the southern United States.
4. The Atlantic storm track may also be affected, moving south and bringing conditions resembling a negative NAO index.
5. El Niño years are associated with the suppression of Atlantic hurricanes.

is almost reversed, with warmer and wetter weather in the Indonesian archipelago and northern Australia and generally cooler and dryer weather in coastal Peru and Ecuador.

Distant effects

By distant effects we mean the effects of El Niño on other regions of the globe that are not in themselves part of the ENSO cycle. Some of these effects are noticeable, especially if the El Niño event is strong, but others are weak and ambiguous and only emerge after averaging over a number of El Niño events to remove the natural variability—or noise—that causes one year to have different weather from another in any case. We first note the overall effect: In an El Niño year, the globally averaged surface temperature can be as much as 0.5°C higher than the years before and after, and this increase is accounted for by the fact that the surface temperatures in the eastern tropical Pacific can be several degrees higher than normal.

One effect that certainly arises during strong El Niño events is that the storm track over the eastern North Pacific becomes stronger, wetter, and further south in winter, bringing heavy rain to central and southern California and additional snow to the southern Sierras. In both the 1982–83 and 1997–98 events, extensive flooding and landslides occurred. In some contrast, further north in Oregon the snow pack was less than usual because of a warm winter but no enhanced precipitation, and Canada tends to experience warmer and drier winters in El Niño years. These effects are not nearly as pronounced in years with weaker El Niño events, and it is often hard to definitively attribute anomalous winter rainfall to an El Niño event.

The storm track does not end when it reaches California, and New Mexico, Arizona, and indeed much of the southern United States can receive enhanced precipitation in El Niño years. (Indeed, New Mexico has some of the best tree-ring records of El Niño.) The storm track may pick up again over the Atlantic and bring enhanced precipitation to Europe. Certainly, there is a correlation between El Niño events and the phase of the North Atlantic Oscillation, with warm events associated with a negative phase of the NAO. Elsewhere in the world, the effects are weaker, but there is some evidence that parts of East Africa (particularly Kenya and Tanzania) experience wetter weather during warm events.

Finally, we mention another tropical effect, a nonlocal one but one that seems to be quite robust. It is that

Atlantic hurricanes tend to be suppressed during El Niño years. Why should this be? Hurricanes form over warm tropical waters, and when the SST in the Atlantic is anomalously high, then the Atlantic hurricane season is more active than usual. However, hurricanes do not respond to the sea-surface temperature alone. Rather, they are a kind of heat engine, and a significant factor in driving hurricanes is the temperature *difference* between the sea surface and the upper atmosphere. (The effect is analogous to the fact that midlatitude weather systems respond to the temperature gradient between the equator and the poles.) During El Niño events, the Pacific region becomes particularly warm, and this warmth spreads throughout the tropics in the upper atmosphere, and in particular the upper tropical atmosphere over the Atlantic becomes anomalously warm. Hence, the temperature difference between the surface and the upper atmosphere that drives Atlantic hurricanes is reduced, and the Atlantic hurricane season in an El Niño year tends to be less active than usual.

7 GLOBAL WARMING AND THE OCEAN

Plurality should not be posited
without necessity.

—Occam's Razor (attributed to
William of Ockham)

IN THIS FINAL CHAPTER WE DISCUSS A TOPIC OF GREAT current and likely future interest, namely *global warming*. In the first half of the chapter, we talk about warming quite generally: what it is, what the evidence is for it, what the consequences might be, and what the level of uncertainty might be about future warming. In the second half of the chapter, we concentrate on the role and effects of and on the ocean. We find that the evidence unambiguously points to a single culprit for global warming, namely the increasing concentration of greenhouse gases in the atmosphere. Although it is possible that we could, with some effort, come up with other explanations that would fit the facts, we would need to invoke several different explanations to fit all the facts and/or invoke rather implausible ones.

GLOBAL WARMING ITSELF

Global warming, in the sense that it is commonly used, is the observed increase in the average temperature of Earth's surface and atmosphere since the late nineteenth century and its projected continuation. The fact that warming has occurred over the past century and is continuing is undeniable. The degree to which it will continue in the future, and the causes of that warming, are both topics of considerable scientific and societal interest. Global warming is, of course, a form of climate change, albeit a forced one and not a natural one. The name "global warming" is useful because it crisply evokes the global nature of the issue: nowhere will be unaffected, nothing will be impervious, no one will be immune. But of course the problem cannot be properly encapsulated by a single global number. The effects will be worse in some places than others, there may be floods here and droughts there, and these regional changes will dictate how society responds or fails to respond to it. Nevertheless, many effects will scale with the change in the globally averaged temperature, so let us initially focus on that and describe the warming that has occurred over the past century or so.

The observed global temperature record

Since the late nineteenth century, the observed average surface temperature has been increasing, as shown in figure 7.1. The data from land comes from about 4,000 stations distributed widely over the the globe, although naturally enough there are more stations in North America and Europe and fewer stations in such places as Antarctica, Greenland, Siberia, and the Sahara Desert. The actual measurements typically are taken twice daily

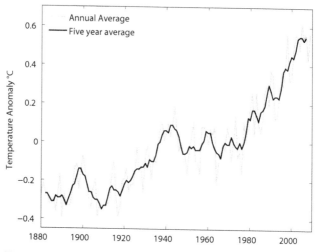

Figure 7.1. The instrumental record of global average surface temperatures from 1880 to 2009, relative to the mean temperature from 1951 to 1980.

(more frequently in some locations) and are of the air temperature a few meters above ground. The ocean data mostly come from in situ observations from ships and buoys; the measurements are the temperature of seawater itself, although in fact this is a good surrogate for the air temperatures just above the surface.[1]

As the temperature data on both the land and the ocean are quite nonuniform, the temperatures are first interpolated onto a regular grid from which a globally averaged temperature can be constructed. (Measurements have in fact been taken on land and on sea since well before 1850, but not with sufficient spatial density to enable the direct construction of a gridded data set.) Obviously, such a procedure is not error-free; errors come from the measurement errors at individual stations, from sampling errors caused by the fact that there may be insufficient coverage over the globe, and from bias errors caused by possible systematic changes in measurement methods. Possible changes in the surroundings of a station, caused for example by urbanization (discussed below), are also a potential source of error. The combined effect of all these errors on the global average is generally estimated to be quite small, less than 0.1°C since 1950, possibly up to about 0.2°C before about 1930, certainly considerably smaller than the increase of temperature over the twentieth century.

There is one possible source of error that has received much discussion both in the scientific literature and in the media, and that is the effect of urbanization. It is common experience that the temperature in a large city is a little higher than that in the surrounding countryside because of all the energy that is expended in the city in the buildings and by transport. Furthermore, the paving over of the ground by asphalt and concrete reduces evaporation and tends to increase surface temperatures. If the stations that measure temperature are in locations that have experienced increased urbanization over the past 50 or 100 years, then any increase in temperature that they have recorded may in part be caused by that urbanization effect, rather than reflecting a true increase in global temperature. One way to study this effect is to compare the increases in temperature between urban areas and areas that have remained rural. When this comparison is done, the effects on the global temperature record are found to be small, contributing an error of about 0.005°C per decade, and about 0.05°C over the past century, although it should be said that some critics of the temperature record believe this error estimate to be too small. Satellite measurements, discussed below, provide another check on urbanization errors.

A somewhat different source of error, although still a bias error and one that has certain similarities with the urbanization problem, is that the way that temperatures have been measured has changed over the past century, both over water and on land. On land the early shelters were fairly heterogeneous; they have been slowly replaced with more standardized shelters known as *Stevenson screens* (after Thomas Stevenson, 1818–1887). These shelters are essentially ventilated white boxes that shield the thermometer against precipitation and direct radiation but that allow air to flow past the thermometer and so give accurate measurements of air temperature. Still more recently, some of these shelters have been replaced with mechanically aspirated shelters to further increase the airflow. The errors due to the different box designs are generally thought to be very small (< 0.1°C) at any given station up to 1950 and negligible after that.

Over water, the methods of taking sea-surface temperature have also varied over time, from taking samples in wooden and then canvas buckets in the early part of the record to measuring the temperatures of the water coming into the engine rooms of ships for cooling. Studies suggest that the errors are small, certainly compared to the changes in temperature over the past century, although changes in measuring techniques have led to some apparently artifactual small jumps in the record.[2]

Satellite measurements

Given that there are possible errors in the direct measurements of temperature, it is useful to compare them with satellite measurements, which provide a completely independent record, less influenced by urbanization issues. Of course, satellites have not been taking measurements for as long as the surface record has existed, and they too are subject to their own errors—difficulties both in calibration and in accounting for the fact that the orbit of a satellite tends to decay over time, potentially contaminating the results.[3] Nevertheless, the combination of satellite and surface observations gives a rather powerful check on temperature increase of the past few decades, as shown in figure 7.2.

Satellite measurements are taken with a microwave sounder, which measures the microwave radiation in several bands, as well as an infrared sounder, which makes similar measurements in the infrared band. The brightness in each band is sensitive to both the temperature and the amount of water vapor in the atmosphere, but by measuring in multiple bands a temperature profile of the atmosphere can be constructed. The temperature trends measured by the satellites agree well with those directly measured at the surface and with those in the lower atmosphere measured by *radiosondes* (instruments carried on weather balloons). The surface measurements show a trend of 0.16°C per decade over the past three decades, and the satellite

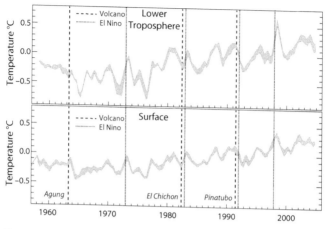

Figure 7.2. Top: Lower troposphere temperature as measured by various satellites and by radiosondes; the gray shading indicates the spread between all measurements. Bottom: Surface temperature records from NOAA, NASA, and UKMO, with gray shading again indicating the spread. Records are monthly means, smoothed with a seven-month running mean filter, and are relative to 1979–1997 mean. Adapted from Solomon et al., 2007.

measurements show trends of between 0.14°C and 0.18°C per decade, depending on the particular method used to obtain temperatures from the brightness measurements.

GLOBAL WARMING IN CONTEXT: THE PAST MILLENNIUM

When discussing global warming, a fair question to ask is, "How does the temperature increase of the past century compare with other periods in the past?" It is a difficult question to answer precisely because we only have direct measurements of surface temperature for the past 140 years or so, although a few individual records go back further; there is a record of measured temperatures in central England from 1659, for example. However, the use of various proxies enables the construction of the temperature record for the past millennium, as shown in figure 7.3.

How are these records constructed? They use proxy data, such as the width of tree rings and the isotopic composition of ice cores, corals, and stalactites. Even more indirect records, such as the time of year of crop harvests and the altitude of the tree line, may also be used where available. These records all provide some measure of temperature. However, to a much greater extent than when dealing with temperatures of the past 100 years, the data coverage is spotty and may reflect only the *local* temperature and in an indirect way, so the difficulty lies in constructing a fair record of *globally averaged* temperature. To construct this record, multiple proxies that are available over the past 100 years or so are combined and calibrated against the instrumental record in such a way that the proxies yield a good approximation of the globally averaged temperature of that period. The multi-proxy record may then be used to reconstruct a globally averaged temperature going back many centuries. The method is accurate to the extent that the proxies have the same relationship to temperature in the past as they do in the present, and to the extent that they can be

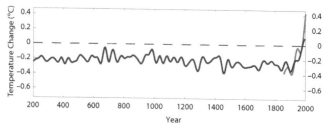

Figure 7.3. Global mean surface temperatures of the past 1,800 years. The lighter solid curve extending from about 1850 to 2000 shows the instrumental record. The longer solid curve is an estimate of temperature over the entire period using proxy reconstructions, and the gray shading is an error estimate (the 95 percent confidence interval). The series are smoothed to remove fluctuations of periods shorter than 40 years, and the temperatures represent anomalies in °C from a late twentieth century value.
Source: Adapted from Jones and Mann (2004).

combined in such a way as to provide a reconstruction of global temperature even though the proxies themselves do not have a uniform global coverage.

Figure 7.3 suggests that the temperatures of the past millennium until about 1900 were relatively uniform compared to the rapid increase in the twentieth century. Note, though, the hint of a "Medieval Warm Period" from about A.D. 900 to A.D. 1200, when temperatures were a little higher than the millennium average, and a Little Ice Age from about A.D. 1400 to A.D. 1800, when temperatures were a little lower. Evidently, though, the rapidity and sustained nature of the warming of the twentieth century was much greater than in any of the previous dozen centuries. The characteristic shape of the curve illustrated in figure 7.3—relatively flat for several centuries and then rising rapidly—has led to it being known as the hockey stick. Although the methodology and accuracy of the calculations leading to the hockey stick have been criticized and continue to be discussed, there remains a consensus among the scientific community, at least insofar as it is represented by Intergovernmental Panel on Climate Change (IPCC) reports, that the temperature reconstructions shown in figure 7.3 are broadly correct. That is, the sustained rate of increase of temperature over the past century is unprecedented, and the current temperature levels are the highest of the past millennium.[4]

FOSSIL FUELS AND GREENHOUSE GASES

Let us now start to look at one potential cause of the observed warming, namely the level of greenhouse gases in the atmosphere. We discussed in chapter 1 that the presence of such gases maintains the surface temperature at a level higher that it otherwise would be: the current average surface temperature is about 15°C, whereas the radiative equilibrium temperature, which would hold if Earth had no atmosphere, is about −18°C. Thus, a rather obvious possible cause of the observed increase in temperatures is an increase in the concentration of greenhouse gases in the atmosphere, and such an increase has indeed occurred.

The main greenhouse gas in the atmosphere is water vapor; however, we do not control its concentration. Rather, as we discussed in chapter 1, there is balance between evaporation (mainly from the oceans, but also from lakes and moist land areas) and precipitation, and the overall level of water vapor is largely determined by the temperature of the atmosphere. The second most important greenhouse gas is carbon dioxide (CO_2), which is an unavoidable product of the combustion of fossil fuels. It has been directly measured at Mauna Loa Observatory in Hawaii for about 50 years, as shown in figure 7.4. The measurements were first taken by C. D. Keeling, and the curve, now one of the most famous in science, is known as the Keeling curve.[5] There are two features about the plot worth noting. The most obvious is the steady increase in CO_2 level over the past few decades, from a level of about 310 ppm (parts per million) to about 390 ppm in 2010. The second is the presence of an annual cycle. There is more land and more vegetation in the Northern Hemisphere, and land takes up CO_2 and releases oxygen during the boreal (Northern Hemisphere) summer, so that there is a minimum of CO_2 at the end of summer (in October), and a maximum in May.

There are two main ways that we know that the increase in CO_2 is caused by the burning of fossil fuels. The first, and most obvious, is that we know how much fossil fuels (mostly petroleum and coal, followed by natural gas) have been extracted from Earth and burned, and from this amount we can calculate approximately how much CO_2 has been added to the atmosphere. Human carbon emissions from fossil fuel combustion (coal, petroleum, and gas) currently contribute almost 9 gigatons of carbon a year to the atmosphere. In fact, more fossil

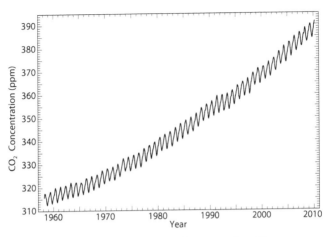

Figure 7.4. The levels of CO_2 measured at Mauna Loa Observatory in Hawaii from 1958 to 2010 in parts per million by volume.

fuel has been burned than can be accounted for by the increase in the atmosphere, and most of the remainder is taken up by the ocean. The second way is via the isotopic composition: the burning of fossil fuels leaves a characteristic fingerprint in the atmosphere. There are three naturally occurring isotopes of carbon, namely, carbon 12, carbon 13, and carbon 14. Carbon 12 is by far the most abundant. Now, carbon 14 is produced in the atmosphere by cosmic rays; it is then incorporated into CO_2 and taken up by plants during photosynthesis. However, carbon 14 is radioactive and decays with a half-life of about 5,700 years. Thus, since the fossil fuels are remnants of plant matter millions of years old, they contain virtually no carbon 14, as it has all decayed away. When fossil fuels are burned, they therefore put CO_2 into the atmosphere that has a *lower* abundance of carbon 14 than the CO_2 already in the atmosphere, and it is indeed found that the ratio of carbon 14 in the atmosphere, relative to the other isotopes, is decreasing at about the right rate to be explained by fossil fuel burning.

Although a continuous record of CO_2 levels from direct measurements goes back only about 50 years, we have a good record of carbon dioxide going much further back, mainly from measurements of CO_2 trapped in bubbles in ice cores in Greenland and Antarctica. There are also a number of somewhat isolated measurements at various periods in the past; for example, a series of measurements were made near Paris from 1876 to 1910. The ice cores reveal that CO_2 levels were about 200 ppm at the last glacial maximum some 20,000 years ago, rising over the course of deglaciation to about 260–270 ppm 10,000 years ago, then slowly rising again to about 280 ppm in 1750, just before the Industrial Revolution.[6] The rate of increase since then, to 390 ppm today, is thus far faster than anything else in the past 10,000 years. The level of methane, another greenhouse gas, has also been increasing over the past few decades, although it has leveled off in the past decade.

POSSIBLE CAUSES OF GLOBAL WARMING

The likely culprit

The atmosphere contains greenhouse gases that absorb and re-emit longwave (infrared) radiation, thus warming the surface. If the greenhouse gas concentration increases, we can therefore expect temperatures to rise. Thus, a reasonable hypothesis is that the increase in anthropogenic greenhouse gases, and in particular CO_2, over the past several decades is responsible for the increase in temperature. In chapter 1 we constructed a very simple mathematical model of this process; we now present a somewhat more refined argument that, although having a similar basis, shows how the vertical profile of temperature plays a role.

Referring to figure 7.5, suppose the temperature profile is initially that labelled "Before." Then let us suppose that we add some greenhouse gases to the atmosphere, increasing its emissivity and absorptivity. Now, the total outgoing longwave radiation to space must remain the same because this radiation balances the incoming solar radiation (which of course stays virtually the same). However, because the absorptivity of the atmosphere has increased, the outgoing longwave radiation, on average, originates from a *higher* level in the atmosphere (radiation emitted from lower levels is reabsorbed by the atmosphere). But the *temperature* at which the outgoing radiation originates must stay virtually the same so that total outgoing longwave radiation stays the same. Thus, the temperature of the atmosphere at the new level where it is emitting radiation to space must increase and, unless the vertical profile of temperature changes significantly (which is unlikely), the surface temperature must increase, as in figure 7.5.

The argument above illustrates an important aspect of the greenhouse effect, which is that an increase in CO_2

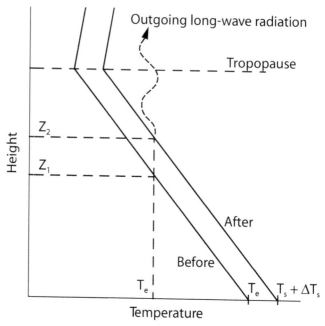

Figure 7.5. Schematic of temperature profiles before and after the addition of greenhouse gases. The total outgoing longwave radiation must remain the same because this radiation balances the incoming solar radiation, and so the emissions temperature, T_e, stays the same. However, the emissions height must increase (from Z_1 to Z_2) because of the increased absorptivity of the atmosphere. Hence, if the temperature gradient in the vertical remains similar, the surface temperature must increase.

levels can have an important effect even if levels of water vapor are already quite high. In this case, it might be thought that since the lower atmosphere is already quite opaque to longwave radiation, the addition of CO_2 would have little effect. However, even in a warm, wet climate the upper atmosphere is quite dry, so that the addition of CO_2 still adds to the emissivity of the upper atmosphere and so raises the emissions height. Because the temperature decreases with height, the overall temperature increases, as in figure 7.5. The question, of course, is how much? That is to say, is the measured increase in greenhouse gases actually responsible for the observed increase in temperature of just under 1°C over the past century?

A partial answer to this question is that the increase that has been observed is entirely and quantitatively consistent with the increase in greenhouse gases. Radiative calculations indicate that the increased downward infrared radiation, ΔR, varies logarithmically with CO_2 concentration, C, according to $\Delta R \approx 5.3 \ln(C/C_0)$, where C_0 is the preindustrial CO_2 level, excluding water vapor and other feedbacks (Myhre et al. 1998). Thus, the radiative forcing increases about 3.7 W/m^2 for each doubling of CO_2, and this rate is sufficient to cause the temperature increase observed. More complete calculations using comprehensive general circulation models of the climate show good quantitative agreement with the observed temperature rise when proper account is taken of greenhouse gases and other anthropogenic influences on climate (such as the fact that we are also putting particulate matter, or aerosols, into the atmosphere), but poor agreement otherwise. A fair question to ask is, if greenhouse gases have been increasing monotonically over the past century, why did the temperature not rise over the middle of the century, from about 1940 to 1975? Most likely, the flat temperature record in midcentury arose because of an increased aerosol concentration in the atmosphere from volcanoes and pollutants, increasing Earth's albedo, possibly in conjunction with natural variability. Regarding the most recent warming, the decade 2000–2009 was 0.2°C warmer than the previous one and about 0.4°C above the 1961–90 average. It was the warmest decade since 1880, and most likely the warmest of the past millennium.[7] The year 1998 was warmed to the tune of about 0.5°C by El Niño, and there have been no comparable El Niño years since, but 2005 and 2010 were nevertheless virtually as warm.

Red herrings and straw men

Of course the models might be wrong and the temperature increase might come from natural causes—natural variability in the climate system, such as we discussed in chapter 6. Thus, the observed warming might not be anthropogenic but rather might be related to our emergence from the Little Ice Age over the past century. This idea is not so much an argument as a speculation because no viable mechanism has been posited that could cause the warming seen over the course of the past century, except possibly for the influence of the ocean, a topic we deal with in the next section. We might imagine that for some reason the ice sheets have retreated, decreasing the albedo and warming the planet, or we might imagine that the clouds have changed configuration in such a way as to cause warming, but there is no evidence for either possibility. If the climate system were so sensitive that such changes were possibilities, we might expect to see evidence of similar changes in past climates. The climate has varied, of course, but looking back at figure 7.3, we see that over the past 1,000 years it has never varied in the same manner as it has in the past 100—the rate of increase of temperature is, so far as we can tell, unprecedented.

One oft-mentioned possibility is that global warming arises because of variations in the sun's output. On the decadal and centennial timescales, variations in the sun's output occur mainly through the solar cycle, a cycle of solar magnetic activity that affects sunspots. The cycle has main periods of about 11 and 22 years, and the former period modulates, albeit slightly, the total solar flux coming into Earth's atmosphere, mainly at ultraviolet wavelengths. The cycle itself cannot cause global warming, but it is possible that there may be longer term variations in the sun's output that modulate the solar cycle, and it is sometimes hypothesized that the Little Ice Age might have been caused by such variations. Indeed, there was a period from about 1645 to 1715, known as the Maunder Minimum, when sunspots seem to have been exceedingly rare, and this period coincided with low temperatures in the middle of the Little Ice Age. Although there is some uncertainty, solar irradiance during the Maunder Minimum is believed to have been about 0.2% less than that of today, or about 0.7 W m^{-2} less.[8] (The change in the ultraviolet component is larger, about 0.7%, but is a small fraction of the total.) The increase in solar irradiance since 1750 is estimated to be at most 0.3 W m^{-2}, and the change since 1900, much less than that. The total solar radiative change is thus considerably less than the changes in radiative forcing that have arisen through changes in the levels of greenhouse gases—a 50% increase of CO_2, for example, is roughly equivalent to a forcing of 2 $W\,m^{-2}$. Unless there are mechanisms in the atmosphere of which we are totally unaware, perhaps somehow amplifying the changes in the ultraviolet component, variations in the solar output are almost certainly insufficient to have caused the bulk of the global warming that we have seen over the past century. If the climate system were so sensitive to variations in the solar cycle, we would expect to have seen a more pronounced variation across the Maunder Minimum.

Finally, we might plausibly ask, could causality of the greenhouse gas–temperature relationship be reversed—that is, might the recent warming be causing the carbon dioxide increase? First, for this to be the case our comprehensive climate models would have to be completely wrong, but that should certainly not be wholly discounted. Second, and perhaps more compelling, no known mechanism could give rise to this reversal of causality on the timescales of decades on which we have seen the warming take place. There are some suggestions in the paleo-record that changes in temperature have preceded changes in carbon dioxide in past ice ages. However, these changes occur on much longer timescales and are likely the result of a positive feedback in which temperature changes cause changes in the ocean circulation, which affect the absorption of carbon dioxide, which changes the temperature, and so on (although the precise mechanism of this sequence of events remains a mystery). *By far* the simplest and most compelling explanation of the current warming is that it is caused by the increase in greenhouse gas concentration.

So what does the future hold, and how certain can we be about it? Before we can answer that, we need to understand the role of the ocean in global warming, so let's turn to that.

HAS THE OCEAN WARMED?

Let us begin our discussion of the ocean's role by asking a simple question: Has the ocean interior itself warmed? The ocean itself has in fact warmed over the years, as shown in figure 7.6, which shows an increase in the heat content from 1955. (There are not sufficient data before this time to extend the time series further into the past.) The heat content is defined as the heat capacity of seawater (which is almost a constant) multiplied by the change in temperature, integrated over the entire mass of the world's oceans. Most of the increase is associated with an increase in temperature in the upper several hundred meters of the ocean; measurements in abyssal ocean are much sparser, although what data there are show a similar but smaller trend, with about two-thirds of the total increase in heat content occurring in the upper 700 m (in spite of the fact that the ocean is on average almost 4,000 m deep).

The ordinate of the top panel of figure 7.6 is the total heat content (i.e., the internal energy), which is perhaps not a meaningful number, so let us make a translation.

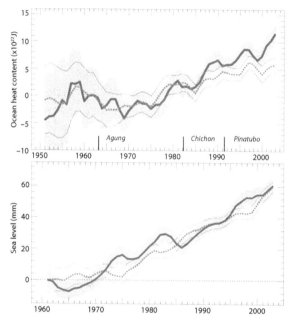

Figure 7.6. Top: The global heat content for the upper 700 m of the ocean (solidblack line, with shading indicating uncertainty) and upper 100 m of the ocean (dotted line, with thin solid lines indicating uncertainty). Bottom: Increase in sea level as estimated from direct measurements (black line and shading) and by combining the contributing components (dotted line and thin solid lines). The time series are all relative to 1961 and smoothed with a three-year running average. Source: Adapted from Domingues et al., 2008.

The total heat capacity of the upper 700 m of the entire ocean is about 10^{24} JK^{-1} (the area of the ocean is approximately 3.6×10^{14} m^2, and the heat capacity of water is about 4×10^3 J kg^{-1} K^{-1}), so that an increase in heat content over the period 1960 to 2000 of 15×10^{22} J corresponds to a temperature increase of about 0.15°C over this period. The rate of increase is obviously not smooth (in fact, some recent estimates suggest that heat content went down in the few years following 2003), and these fluctuations are likely caused by some form of natural variability in the climate system. Nonetheless, the overall upward trend is unmistakable.

Implications for global warming

Earlier we noted that one possible cause of the warming over the past century is natural variability, with one candidate mechanism being a variation in the ocean that releases heat to the atmosphere, thus causing atmospheric warming. This is a very plausible conjecture because the ocean certainly has the potential to vary on the timescale of decades to centuries; however, the empirical evidence is against it. If the ocean were to have given up its heat to the atmosphere, then it would have become *cooler* over the past few decades. In fact, the heat content of the ocean has *increased* over that period, as we saw in figure 7.6. This increase implies that global warming is not coming *from* the ocean, for this would cause the ocean to cool; rather, the ocean is warming *because of* global warming.

A temperature increase of 0.15°C is fairly small, much less than the increase in surface temperature of about 0.6°C that occurred over the second half of the twentieth century. The temperature increase of the abyssal ocean is likely to be even smaller, although observations are sparse. The small increase is important because it tells us that global warming still has a long way to go. Eventually, except for possible effects caused by changes in the ocean circulation, we expect that the increase in temperature of the deep ocean will be similar to that at the surface. Even if we were to stop adding greenhouse gases to the atmosphere today, it would take a very long time for the deep ocean temperature to rise to that level, and until that time, the surface temperature will keep rising, as we discuss later in this chapter.

More Effects of and on the Ocean

Let's now consider how the ocean might affect global warming and how global warming might affect the ocean. One of the most important effects stems from the fact that global warming is a form of climate variability, albeit variability of a forced kind; thus, just as with natural variability, the ocean damps the response in some fashion. There are other potential effects, so let's make a short list of them and then consider them one by one.

1. The damping effect will slow down the warming and delay the time it takes to reach a new equilibrium state.
2. The ocean circulation may change in some fashion. The most worrisome possibility is that the overturning circulation will change substantially, and possibly even shut down.
3. The sea level will rise, both because of an expansion of seawater as it warms and because of the melting of land ice, such as glaciers.

4. The ocean will absorb some of the CO_2 that is added to the atmosphere. he extent to which it does will obviously affect the rate and ultimate level of warming. It will also lead to changes in ocean acidification and possible changes in ocean biology.

Each of these effects is quite important and we devote a section to each, except for the last one, which is beyond our purview. However, we do note that it may take many centuries for the ocean or any other component of the climate system to draw down anthropogenically enhanced levels of CO_2 in the atmosphere.[9]

THE SLOWING OF GLOBAL WARMING

Perhaps the most important and least ambiguous effect of the ocean is that it acts to slow down global warming in a certain sense, although the effect is rather subtle. Furthermore, the ocean is not likely to significantly affect the final equilibrium temperature that the planet will reach if, let us say, CO_2 levels eventually double before leveling off. So just what does the ocean do? Let's try to explain it first just using words; appendix A to this chapter provides a mathematical treatment of the same issue.

In chapter 4 we noted that in the upper ocean there is a mixed layer, typically 50–100 m thick, in which the vertical distribution of temperature and salinity is almost uniform. This turbulent region is stirred by the winds and convection, and its heat capacity is about twenty times that of the atmosphere, which means that it responds rather slowly to such things as daily changes in the weather. However, compared to the deep ocean, it responds very quickly. Thus, for example, if the radiative forcing were to change because of global warming, the mixed layer might be expected to respond and come to a new equilibrium on the timescale of a few years. Now, the radiative forcing has been changing rather slowly over the past century or so, and the mixed layer has little difficulty keeping up. At most, the mixed layer is in equilibrium with the forcing levels in the atmosphere a decade ago, and likely just a few short years ago. The atmosphere, which has a much smaller heat capacity than the mixed layer, tends to respond to the mixed layer quickly. Thus, increased radiative forcing causes the ocean's mixed layer to warm quickly, and this warming in turn sets the atmospheric temperature.

However, the rest of the ocean takes longer—*much* longer—to equilibrate. To get a rough sense of how much longer this time might be, note that in round numbers the mixed layer is 50 m deep, the thermocline is 500 m deep, and the entire ocean, 5,000 m deep; roughly, the times scale accordingly. What effect might these differences have on global warming? Let us suppose that we add some greenhouse gases to the atmosphere and so increase the downward flux of radiation at the surface. Over a few years, the mixed layer warms up until it reaches an equilibrium—that is, a state in which it gives up as much heat as it is receiving—although the temperature of the deep ocean will hardly have risen at all over that period. However, the increased heat that the mixed layer is giving up is going, in part, to warm the ocean below, and it takes a long time, perhaps many centuries, for the deep ocean to fully equilibrate because it is so big. As the deep ocean warms, the mixed layer can give up less of its heat to the ocean below, and so can only balance the radiative forcing by further increasing its temperature, so that it gives its heat back to the atmosphere.

What is the consequence of this picture for global warming? We are slowly but steadily putting greenhouse gases into the atmosphere; the mixed layer responds to this input, and its temperature increases in concert. However, the deep ocean is *far* from equilibrium, which means that, even if we were to stop adding greenhouse gases to the atmosphere and the levels of greenhouse gases were to stabilize at some level, the temperature of the ocean's mixed layer, and so of the atmosphere, would continue to rise for a long period after that. Let us suppose that we continue putting CO_2 into the atmosphere until its level has doubled from that in preindustrial times, and that this doubling occurs in the middle of the twenty-first century. We can expect the global averaged temperature to rise by between 1.3°C and 2.5°C, and probably around 1.8°C, from its preindustrial value by then.[10] Suppose that at that time the political and technological stars align and we are able to prevent greenhouse gas levels in the atmosphere from increasing any further. The average surface temperature of Earth will nevertheless *keep on increasing* until the deep ocean has finally equilibrated, which will take an additional few hundred years or more.

There is considerable uncertainty as to what this final equilibrium temperature rise will be, but it may be much higher than the 1.8°C mentioned above; most estimates are between 2°C and 4.5°C, although higher values cannot be definitively excluded. If we were to eventually cease CO_2 emissions entirely, perhaps because fossil fuels run out, it would take centuries for the CO_2 to finally revert to levels near or a little above the preindustrial value (Archer 2010). During that period, temperatures would likely stay roughly constant for a few centuries before slowly falling back down to levels commensurate with the level of CO_2, as illustrated in figure 7.7. If CO_2 remains doubled for a century before emissions cease, the peak warming will probably be a little over 2°C, and if it triples the peak warming will be around 3°C, with uncertainties of about plus or minus 0.5°C. These peak levels will remain for centuries. Whichever way we slice it, global warming is a long-term problem.

CIRCULATION CHANGES AND A THERMOHALINE SHUTDOWN

One possibility that is raised from time to time is that global warming will bring about a slowdown or cessation of the meridional overturning circulation, sometimes called a *thermohaline shutdown*.[11] Is such a thing likely? First recall that in chapter 4 we divided the ocean circulation up into a quasi-horizontal circulation that is primarily wind driven (the gyres) and an overturning circulation. With warming, the winds may change in

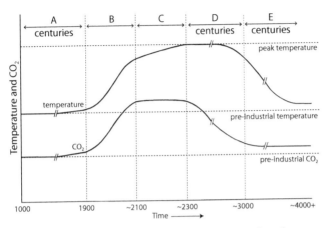

Figure 7.7. Schema of a CO_2-temperature scenario. Carbon dioxide levels increase from 1900 to 2100 (period B) before leveling off
(period C) because of controls on emissions. Temperature increases
rapidly in period B, then more slowly in period C. At the end of
period C (the year 2300 in the figure), anthropogenic emissions go to
zero, and the level of CO_2 slowly diminishes through periods D and
E back to levels close to, but probably a little above, the preindustrial
period. In period D, temperature stays roughly constant for centuries before it too eventually falls back to near pre-industrial levels in
period E. Many plausible scenarios can be adapted from this plot by
changing 2100 and 2300 to other dates and calibrating the y-axis.

detail but almost certainly will not change in their basic
structure. Thus, we can confidently predict that the gyres
and their western boundary currents will remain qualitatively unchanged in the decades ahead, although certainly there might be shifts in latitude of a few degrees if
the surface winds were to change by that amount.

The overturning circulation could, conceivably, change
by a larger amount. Recall that one controlling factor in

the intensity of the overturning circulation is the meridional buoyancy gradient at the surface of the ocean. Convection at high latitudes in the North Atlantic occurs because the surface waters there are sufficiently dense that they sink in convective plumes, and an overturning circulation ensues. Two things could happen to bring this process to a halt. First, if there were to be increased rainfall or a release of freshwater from the melting of glaciers at high latitudes, the surface waters of the ocean will freshen and become lighter, and thus become less likely to sink. Second, if the surface waters were to become warmer, then they would become lighter, and this too could lead to a cessation in the overturning circulation. Note that this latter effect will only occur when global warming has proceeded far enough to make a sufficient difference to the surface density at high latitudes. However, we would not expect it to be a permanent effect because ultimately the overturning circulation is maintained by a *difference* in density between high latitudes and low latitudes, and if the primary effect of global warming is simply to raise temperatures worldwide, these differences are eventually likely to be similar to what they are today. If there are permanent changes in precipitation patterns, however, the circulation could become permanently weaker if the high-latitude ocean stays fresher.

We cannot tell with any certainty whether a shutdown in the overturning circulation will occur, but recent experiments with comprehensive climate models suggest that, at least over the next several decades, a complete shutdown is quite unlikely. However, a number of climate models do show a weakening of the overturning circulation by a significant fraction, mainly caused by increased high-latitude rainfall, and then an eventual recovery. The reductions vary considerably from model to model, from no reduction to a reduction of about 50%, with an average of about 25% reduction. The worst-case scenario would be a melting of part or all of the Greenland ice sheets, bringing a surge of freshwater to the North Atlantic that might well cause overturning circulation to halt, but that is unlikely in the near or medium future.

The consequences of a severe slowdown in the overturning circulation could be quite drastic because it would mean that the ocean would transport less heat to high latitudes and, ironically, this could cause temperatures at high latitudes to fall under global warming. This change would, ultimately, most likely be a self-correcting consequence because a fall in temperature at high latitudes would create conditions conducive to the overturning circulation turning back on. In fact, however, the reductions in the overturning circulation that models predict for the twenty-first century are sufficiently mild that even in and around the North Atlantic Ocean, including northwestern Europe and Greenland, where the effects are likely to most noticeable, the overall increase in temperature caused by global warming outweighs the cooling caused by the reduced ocean heat transport.

If global warming were to continue into the twenty-second century and beyond and produce a warming of several degrees Celsius, then all bets are off. It is possible that the deep meridional overturning circulation would significantly weaken and possibly even fall to zero for several decades, and possibly centuries, until the deep ocean temperature warms and the meridional circulation re-establishes itself in the new climate equilibrium. The timescales for the overturning circulation to turn off and back on again are measured in decades and centuries, and if there really were no significant meridional overturning circulation for such a period, the consequences could be severe indeed, with possible wholesale changes in climate at high latitudes. It should be said that most climate models do *not* predict a shutdown in the foreseeable future, but the consequences could be severe if a shutdown were to happen. Global warming makes society confront possibilities that are unlikely but, if they do happen, would produce severe consequences.

SEA-LEVEL RISE

An almost certain consequence of global warming on the ocean is that sea level will rise, if only because as water warms it expands. In the oceans, the only way that an increased volume of the ocean can be accommodated is by an increase in sea level, and sea level has indeed risen over the past several decades, as illustrated in figure 7.6. Sea level is estimated to have risen about 20 cm since records began in the late nineteenth century, and it rose at about 2 mm per year over the last half of the twentieth century, increasing to about 3 mm per year from 1993 to 2003. The fact that sea level has increased over the past century is, by the way, additional evidence that global warming is arising from an external cause, and not from internal variability involving the ocean giving up heat to the atmosphere. If the latter were the cause, then the ocean would be cooling and sea level would be falling (except possibly because of the effects of ice melt, but this is insufficient by itself to cause sea level to rise as observed).

Let us do a basic calculation to see how much sea level can be expected to rise from thermal expansion alone. From the equation of state for seawater, namely equation 2.1, the volume of the oceans will change with temperature according to

$$\frac{\Delta V}{V} = \beta_T \Delta T \tag{7.1}$$

where ΔV is the change in volume, V is the current volume, T is the temperature, and β_T is the coefficient of thermal expansion. This coefficient itself changes with temperature and pressure from an average value of about $2.4 \times 10^4 \, \text{K}^{-1}$ at the surface in low latitudes to about $1.3 \times 10^4 \, \text{K}^{-1}$ for the ocean below the thermocline (which of course is most of the ocean). Most of the increase in ocean volume will be accommodated by an increase in the depth of the ocean, and given that the ocean is on average 3.7 km deep and using $\beta_T = 1.4 \times 10^4 \, \text{K}^{-1}$, we find that an increase in temperature of 1°C will result in an increase in sea level of, approximately, half a meter.

This is a significant overestimate of the rate of thermal expansion that has happened over the past century, or what is likely to happen in the near or medium-term future because the temperature of the deep ocean has changed only a little. Most of the temperature increase thus far has been concentrated in the upper ocean: in the mixed layer and in and above the main thermocline. If we suppose that the depth of the column of water experiencing this temperature increase is just 1 km, and using $\beta_T = 2.0 \times 10^4 \, \mathrm{K}^{-1}$ (appropriate for the warm upper ocean), we obtain a rough estimate of 20 cm of sea-level rise for each degree rise in temperature. As time progresses and the deep ocean begins to participate in the temperature rise, this rate can be expected to increase to the aforementioned 50 cm per degree, but that will only occur some time in the future. Still, in the long term, such an increase is unavoidable if global warming continues. If the equilibrium climate response to a doubling of CO_2 is 3°C, which is a typical estimate, and if temperatures reach and stay at that level for an extended period, sea level will eventually rise by somewhere in the vicinity of 1.5 m purely by thermal expansion, without any contribution from the melting of ice sheets.

Needless to say, there are other possible causes of sea-level rise as well as thermal expansion, the most important one being the decrease in ice over land (glaciers and ice sheets), but not, of course, the melt of ice floating on seawater. In the past few decades, thermal expansion of the oceans is estimated to have contributed a little more than half of the total sea-level rise, glaciers and ice caps a little less than a third; losses from the polar ice sheets contributed the remainder, with total sea level currently rising at about 2.4 mm per year (Domingues et al. 2008 and Bindoff et al. 2007). For the future, calculations using comprehensive climate models project that a global sea-level rise due mainly to thermal expansion (projected to be about two-thirds or more of the total) of a few tens of centimeters might be expected in the coming century if greenhouse gas emissions continue, but there is quite of lot of uncertainty and, of course, it depends on the rate of greenhouse warming itself. Such increases in sea level would mainly affect very low-lying communities such as the Maldives, in the Indian Ocean, and Bangladesh. It has been said that about half of Bangladesh would be flooded if the sea level were to rise by 1 m, creating millions of refugees. Even with a sea-level rise of 30 cm, the consequences would be severe.

Much higher rises in sea level would ensue if the great ice sheets covering Greenland and Antarctica were to melt. The ice on Greenland (about 2.8 million km^3) would then contribute about 7 m to sea-level rise, the West Antarctic ice sheet (about 2.2 million km^3), about 5 m. The East Antarctic ice sheet is the biggest of them all (about 28 million km^3), and if it were to melt, sea level would increase by about 70 m, but such an increase is extremely unlikely, at least in the foreseeable future. It is also generally thought unlikely that there will be wholesale melting of the Greenland or West Antarctic ice sheets in the twenty-first century, even if global warming continues apace, if only because it takes a long time for the ice to melt and flow into the ocean. Nevertheless, our knowledge of the dynamics of land ice sheets remains rudimentary and there does remain the possibility that we have underestimated the risk of significant melting of land ice (glaciers as well as the big ice sheets), and hence underestimated the possibility of significant sea-level rise. Even if sea-level rise is modest in the coming decades, on timescales of more than 100 years the possibility of significant land-ice melt and of sea-level rise of much greater than 2 m should not be discounted.[12] Rather interestingly, many climate models predict that the Antarctic ice sheet will experience increased snow fall in the coming decades (because climate is warmer so there is more moisture in the atmosphere) but will *not* experience additional melting (because it will still be too cold). Thus, the ice sheet may grow slightly, causing sea level to fall. In any case, the extent and pace of ice sheet melting is currently one of the more significant unknowns in the global warming business.

LOSS OF SEA ICE

Sea ice is formed by the freezing of seawater and so is to be found at high latitudes in both the Northern and Southern hemispheres, especially in their respective winters and springs. Sea ice is important to the climate system because it has a higher albedo than seawater (and therefore reflects back more solar radiation to space), because it affects the way the atmosphere and ocean exchange heat and water vapor, and because when sea ice forms, salt is extruded into the ocean, and when it melts, freshwater is added to the ocean. As the oceans warm, it would seem natural for the extent of sea ice to diminish, and indeed there is evidence (both anecdotal and quantitative) that sea ice already has diminished. If the sea-ice extent were to continue to diminish, then the consequences could be quite significant, from changes in physical effects already mentioned to the opening of the famed Northwest Passage through the Arctic Ocean for shipping, to diminished habitats for some of the most charismatic life forms on the planet, such as polar bears and penguins. What has happened in the past, and what will happen?

Before the era of satellites, observations of sea ice were taken somewhat intermittently from ships and aircraft and from the coast, but the observations were naturally hard to synthesize into a coherent record of total sea-ice extent. Since the 1970s, satellites that detect passive microwaves emitted from the surface have provided a much more comprehensive record. Objects near the surface emit radiation over a whole range of wavelengths, and although most of it is in the infrared range, microwave radiation is also emitted, albeit at relatively low energy levels. Unlike infrared radiation, microwave emission is determined not so much by the temperature of an object as by its physical composition—its crystalline structure, for example. Of relevance to us, sea ice emits more microwave radiation than does seawater. Furthermore, clouds emit very little microwave radiation, and the radiation emitted from the surface passes through clouds. For these reasons, passive microwave detection is a good way to measure sea ice, year round, day or night, and in cloudy or clear sky conditions.

Over the past couple of decades, sea-ice area has certainly diminished, particularly in the Northern Hemisphere (figure 7.8). The biggest changes have been in the summer months, where the change is about twice that of the annual variation, so the changes have been significant but not overwhelming. The total ice area in the Northern Hemisphere varies between about 6×10^6 km^2 (in summer) and 14×10^6 km^2 (in winter), so the decrease corresponds to about a 10% drop of the summer value. However, in the year 2007, the ice cover fell by about 25% over previous years, with a subsequent partial recovery, so it is possible that the decline in Arctic sea-ice cover is accelerating, but there is insufficient data to be definitive.

Perhaps the biggest fear associated with diminished sea ice is that it will enable the climate system to reach a so-called tipping point, beyond which changes in climate are large and almost irreversible. As ice diminishes, less solar radiation is reflected to space and so more is absorbed at Earth's surface, further warming the climate, and so on. The effect is most pronounced in summer, when sea ice is already at a minimum, because in winter the amount of incident solar radiation is already small. It is hard to be definitive about whether, as global warming progresses, such an effect could lead to a rapid diminution and even total disappearance of Arctic sea ice (where effects have been largest so far) in summer. Most model calculations do not suggest that a disappearance is likely to happen in the near future, but if the level of CO_2 were, for example, to quadruple (which it easily will if we freely burn all the coal, oil, and shale buried in Earth) and the

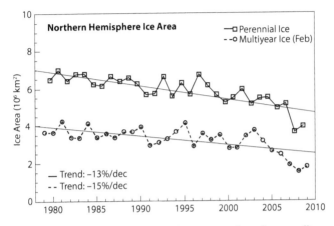

Figure 7.8. Sea ice cover for the Northern Hemisphere from satellite data. Perennial ice excludes seasonal ice cover, and multi-year ice accounts only for ice that has existed for more than one season. Source: Adapted from Comiso, 2002, and Comiso et al., 2008.

climate system were to warm by several degrees Celsius (which would then be quite likely), disappearance cannot be ruled out. Indeed, given that such warming would, as we have discussed, then likely persist for centuries, there is a distinct possibility that the ice sheets on land would also melt, with still more catastrophic consequences.

GLOBAL WARMING— SOME PERSONAL REMARKS

In this last section, I would like to emphasize two aspects about global warming that do not, I think, get sufficient attention: We need to think in terms of probabilities or

GLOBAL WARMING IN A NUTSHELL

Causes and evidence

- Carbon dioxide (CO_2) and a few other gases are greenhouse gases, meaning that they absorb and re-emit longwave radiation that is emitted from Earth's surface, maintaining the surface at a higher temperature than it would be in their absence (about 15°C as opposed to −18°C).
- Greenhouse gases, and in particular CO_2, are added to the atmosphere by the burning of fossil fuels. CO_2 concentration has steadily increased since the beginning of industrialization, from about 270 ppm in 1750 to about 295 ppm in 1900 and 390 ppm in 2010. It will probably exceed 400 ppm some time in 2014.
- The average surface temperature has also increased since preindustrial times and by about 0.8°C over the past century.
- The increase in radiative forcing at the surface is approximately linear—a little less than 4 Wm^{-2} for each doubling of CO_2. This increased forcing (along with other greenhouse gases, and ameliorated by aerosols) can readily account for the observed increase in temperature over the past century.

Projections

- If carbon dioxide levels keep steadily rising and double in the coming century, average temperature increase is expected to be within a range of 1.3°C to 2.5°C, and most likely about 1.8°C, from preindustrial levels by the time of doubling.
- Even if carbon dioxide levels were to stabilize at twice the preindustrial value some time in this century, the temperature would keep on slowly rising to finally reach between about 2°C and 4.5°C higher than the preindustrial value when the ocean equilibrates after a number of centuries.

- If anthropogenic carbon dioxide emissions were completely halted, it would take several centuries for the level of CO_2 in the atmosphere to revert to its preindustrial value. Hence, temperatures would remain high for several centuries.

Other effects on and of the ocean

- Sea level is projected to rise on average by somewhere in the region of 0.4 m over the next century, mostly because of the expansion of seawater as it warms, plus some ice melt, but the uncertainty is large (a factor of 2 either way) and the rise may not be uniform.
- A major melting of the major land-ice sheets is unlikely over the coming century, but if one were to begin, the consequences could eventually be catastrophic, with a sea-level rise of about 6 m if either the Greenland or West Antarctica ice sheets were to completely melt.
- Global warming, once it has occurred, will persist for centuries. Thus, a significant reduction of sea ice and a melting of the Greenland and West Antarctica ice sheets, with concomitant changes in ocean overturning circulation, cannot be ruled out on these timescales. But our ignorance is profound on such matters.

likelihoods, and we need to think clearly about the timescales involved.

Thinking about probabilities is necessary because we don't understand the climate system fully, so we don't know for sure what will happen in the future. The probability we assign to something is then really a measure of the confidence we have in that outcome. However, unlike a weather forecast, it is not as if the climate will warm by either 1°C or 5°C depending on the outcome of some chance weather event, some metaphorical coin toss, that might occur somewhere in the system. Rather, if we build up the level of greenhouse gases to some particular amount, then some definite amount of warming will occur, with 100% certainty. We just don't know what that amount is! We might never know, until the warming actually occurs. Thus, the probability represents our degree of belief in something. It is not just a wholly subjective degree of belief, because it is based on calculations that in turn are based on sound physical principles and the laws of nature, but the probability does represent our lack of knowledge of the system. It is useful to try to be quantitative about our uncertainty, by saying something like "We are 90% sure of something," both because it is useful to be as definite about something as we can possibly be and because there are then ways to calculate the probabilities of other uncertain events occurring, using various statistical procedures. Still, it is good to be somewhat skeptical about such quantitative measures when they primarily reflect our own ignorance.

A particularly disturbing aspect of global warming is that there is a small likelihood that something catastrophic will happen—that the Greenland ice sheets will melt and raise the sea level by several meters, for example. We currently believe that the chances of this are very small, at least in this century, but the consequences would be catastrophic to many countries of the world. I personally would not cross a street if the chances of being run over were one in a thousand. The chances might be small, but the consequences, at least to me, would be large. Whether we should live with the risk of potentially catastrophic global warming or "take insurance" by trying to curb emissions today is a question for society as a whole. (Some degree of global warming is of course inevitable.) When dealing with risk, we have to take into account both the likelihood of something happening and the consequences if that something does happen, and we need to weigh the overall risk against the cost of taking insurance. We "buy" insurance by investing in alternative sources of energy, renewable and nuclear, and by living, where possible, less wastefully. There is little downside to this for the developed nations: the cost is not prohibitive compared to the consequences, and in the worst case we prepare for global warming by being efficient and environmentally sound and then the warming turns out to be less than anticipated. For the undeveloped nations and emerging economies such as China and India, the transition to an economy less dependent on fossil fuels may be far more problematic.

The second aspect is the one of timescales, and this is where the ocean in particular comes in. Over the next several decades it is quite plausible (although take heed of the previous paragraphs!) that global warming will continue at about the pace we have seen in the last century. If, let us say, carbon dioxide levels increase in the atmosphere at about 1% per year, then in somewhat less than half a century, they will reach double their preindustrial level and we might expect temperatures to further increase by somewhere around a degree, for a total increase in a range of 1.3°C–2.5°C over the pre-industrial value, perhaps most likely about 1.8°C (this figure is known as the transient climate response). Let us imagine that by this time technology has improved sufficiently that emissions can be considerably reduced at little cost to our standard of living and that levels of greenhouse gases in the atmosphere then stabilize. However, as we discussed earlier in this chapter, temperatures will almost certainly keep on rising. And they will do so until the deep ocean finally equilibrates in hundreds of years, or until emissions virtually cease. There is much more uncertainty as to what the value of the *final* equilibrated temperature increase (the equilibrium climate sensitivity) will be than for the transient value, because with larger responses and on long timescales, other effects and feedbacks potentially start to come into play (possible changes in cloudiness, the wholesale melting of the ice sheets, and the positive ice–albedo feedback, for example). We noted earlier that the likely equilibrium response is between 2°C and 4.5°C, but there is some possibility of still higher values.

One might of course hope that once anthropogenic emissions have stabilized there will be a drawdown of CO_2 by the ocean and by vegetation on land. However, calculations of the carbon cycle suggest that levels of carbon dioxide in the atmosphere will not, in fact, significantly diminish for many centuries after human emissions cease. Let us also note that if we do burn all Earth's resources of coal and petroleum (including heavy oil and shale), carbon dioxide levels may well go up by a factor of six or more with a corresponding warming of almost certainly more than 5°C and possibly more than 10°C, staying at that level for centuries and giving the great ice sheets on Greenland and West Antarctica plenty of time to melt.

Given all the above, one scenario of the future is the following. Let us suppose that we continue to burn fossil fuels for the next few decades, but (rather optimistically) let us also suppose we make good efforts to curb emissions and finally succeed in doing so a few decades hence, and that we are able to stabilize the level of greenhouse gases in the atmosphere at about double the preindustrial level. On this timescale (the short timescale, given the nature of the problem), the additional global warming will likely be a degree or so Celsius, and although there will be some significant regional changes (perhaps especially in precipitation and in extremes of climate), climate change overall may well be less dramatic than the dire scenarios that are sometimes portrayed in the media. However, in the longer term (several decades to centuries and beyond) the problem may be worse that is often expected because global warming will continue relentlessly with consequences to match. An eventual 3°C rise in temperature, which is quite likely if carbon dioxide levels modestly double and stay doubled, will have very large consequences if and when it persists for centuries. Eventually, of course, emissions will diminish or cease if only because fossil fuels run out or become uneconomical, and a scenario for that is illustrated in figure 7.7. If CO_2 levels were to increase to double (or triple) the present value and emissions were, optimistically, to completely cease a century or two after that, the temperature would remain at more than 2°C (or 3°C for tripling) above the preindustrial value for several centuries.

Set against this bleak scenario is the likelihood that society itself will evolve in unforeseen ways, potentially making our current fears moot. Perhaps, then, we simply should not plan for the long term? Perhaps, as Estragon said so memorably in *Waiting for Godot,* there is "Nothing to be done," as the short term will not be so bad and in the long term human development itself is unpredictable. But if we do nothing, then almost certainly carbon dioxide levels will more than double this century and may well double again the following century. Unless we are somehow able to engineer our way out of trouble (for example, by trapping and sequestering the CO_2 emitted when fossil fuels are burned, or extracting CO_2 from the atmosphere), the climate change that will inevitably follow will almost certainly significantly affect the planet Earth itself and all the life on it.

APPENDIX A: MATHEMATICS OF THE TWO-BOX MODEL

Here we give a mathematical description of the two-box model of the ocean, illustrated in figure 7.9. The evolution equations of the two boxes are

$$C_m \frac{dT_m}{dt} = F - \lambda_1 T_m - \lambda_2 (T_m - T_d), \qquad (7.2a)$$

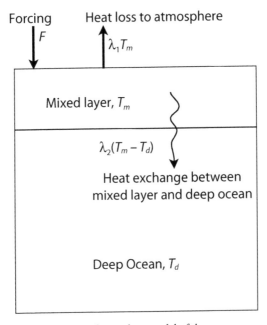

Figure 7.9 A simple two-box model of the ocean, with a mixed layer at a temperature T_m and a deep ocean layer at a temperature T_d, and exchanges of heat between the components as shown.

$$C_d \frac{\mathrm{d}T_d}{\mathrm{d}t} = \lambda_2 (T_m - T_d). \tag{7.2b}$$

In these equations, t is time, T_m and T_d are the temperature anomalies of the mixed layer and deep ocean, respectively, F is the anomalous radiative forcing caused by greenhouse warming, and C_m and C_d are the heat capacities of the mixed layer and deep ocean, respectively. The parameters λ_1 and λ_2 are exchange coefficients that determine the rate at which heat is transferred from the upper ocean to the atmosphere and from the upper ocean to the deep ocean, respectively. Although an exact solution of the above equations is often possible (depending on the form of F), it is more informative, and more general, to look at approximate solutions, and that is how we will proceed. The main assumption we make is that the heat capacity of the deep ocean is far greater than that of the mixed layer ($C_d \gg C_m$), which is a good assumption considering that the depth of the mixed layer is typically $\leq$ 100 m, whereas the depth of the ocean itself is on average about 4,000 m.

Given the big disparity in heat capacities, there will be two timescales to the problem: a short timescale over which the mixed layer comes into a quasi-equilibrium, and a much longer timescale over which the full ocean equilibrates. In the short timescale, the deep ocean does not respond and its temperature stays at the initial temperature, namely zero (because all temperatures are measured relative to the initial temperature). Equation 7.2a becomes

$$C_m \frac{dT_m}{dt} = F - (\lambda_1 + \lambda_2) T_m. \tag{7.3}$$

If we turn on the forcing and then hold it constant, the solution of equation 7.3 is found to be

$$T_m = \frac{F}{\lambda^*} (1 - e^{-t\lambda^*/C_m}), \tag{7.4}$$

where $\lambda^* = \lambda_1 + \lambda_2$. There are two conclusions to be drawn at this stage:

1. The system evolves toward a quasi-equilibrium on a short timescale of $t_s = C_m/\lambda^*$. Observations and experiments with comprehensive climate models suggest that this timescale is on the order of a few years to a decade.
2. The quasi-equilibrium temperature reached on this short timescale is given by $T_2 = 0$ and

$$T_m = \frac{F}{\lambda_1 + \lambda_2}. \tag{7.5}$$

Let us now consider timescales much longer than C_m/λ^*. We suppose that the mixed layer is in a quasi-equilibrium and that equation 7.2 may be approximated by

$$0 = F - \lambda_1 T_m - \lambda_2 (T_m - T_d), \tag{7.6a}$$

$$C_d \frac{dT_d}{dt} = \lambda_2 (T_m - T_d). \tag{7.6b}$$

The mixed-layer temperature is thus given by

$$T_m = \frac{\lambda_2 T_2 + F}{\lambda_1 + \lambda_2}, \tag{7.7}$$

and substituting this in equation 7.6b gives

$$C_d \frac{dT_d}{dt} = \frac{\lambda_2}{\lambda^*} (F - \lambda_1 T_2). \tag{7.8}$$

The system now evolves on the long timescale $t_l = C_d \lambda^* / (\lambda_1 \lambda_2)$, which, given the large value of C_d, may be measured in centuries. The final equilibrium reached has the temperature

$$T_m = T_d = \frac{F}{\lambda_1}, \tag{7.9}$$

which is higher than the temperature given by equation 7.5.

Let us end with few cautionary notes and general remarks. First, and to summarize, there is a fast evolution to the temperature $F/(\lambda_1 + \lambda_2)$, followed by a much slower evolution to the final temperature F/λ_1. However, the real ocean does not consist of just two boxes; the real ocean is immensely complicated. There is indeed a big separation between the timescale on which the mixed layer responds and that on which the full ocean equilibrates, but there are a number of intermediate timescales on which other aspects of the oceans respond, like the gyre and the thermocline. So our treatment is a simplification. Nevertheless, it does contain an essential truth, and that is that it will take a long time for the ocean to equilibrate fully. Even if we were to curtail our emissions of greenhouse gases into the atmosphere and the levels were to stop increasing, the temperature would slowly keep on increasing, possibly for hundreds of years, until the true equilibrium were reached.

Notes

NOTE TO CHAPTER 1

1. The numbers in this table are obtained from Schmidt et al. (2010). See also Myhre et al. (1998) and for a more informal discussion, http://www.realclimate .org/index.php/archives/2005/04/water-vapour -feedback-or-forcing/.

NOTES TO CHAPTER 2

1. If we are speaking about the ocean as a whole, we might say "the ocean" in the singular. If we wish to emphasize that the body of water is composed of several components, we might say "the oceans," plural. But exceptions abound and often either way is fine; it is best not to get hung up about it.

2. Salinity used to be, and sometimes still is, measured in practical salinity units (psu), which are defined in terms of the electrical conductivity of seawater. Conductivity is easy to measure and a psu is almost equal to a part per thousand, but the psu is not a proper unit of concentration and is awkward to incorporate into an equation of state, and so mass concentration itself is a better unit from a more fundamental point of view.

3. The figure actually shows potential density, which, roughly speaking, is the density with a small correction for the compression by pressure. The data are from the National Oceanographic Data Center's World Ocean Atlas.

NOTE TO CHAPTER 3

1. A more detailed but elementary account of the Coriolis force is given by Persson (1998). More mathematical treatments are given in any number of more advanced textbooks, including Pedlosky (1987) and Vallis (2006). In fact, a mathematical treatment can be far simpler and more straightforward than that given here, provided the student is comfortable with a vector cross-product.

NOTES TO CHAPTER 4

1. The question of what "drives" the ocean has led to some confusion and disagreement in the scientific literature, in part because the term "drive" means different things in different contexts. A car is driven (i.e., controlled) by a driver, usually human, but the car wheels are driven (i.e., powered) by the engine. Unfortunately, in science, too, the word is sometimes used to mean powered by, and sometimes controlled. To avoid the word completely would be tilting at windmills, but we try to use it only in contexts in which its meaning is unambiguous.

2. The torques that we refer to in this section are local effects, enabling a parcel to spin like a propellor, albeit thousands of times slower. Technically, they are the curls

of (i.e, $\nabla\times$) the forces—"force-curls"—rather than the force times the distance from a fulcrum. The pressure gradient force does not provide a torque; its curl is zero: $\nabla \times \nabla p = 0$.

NOTE TO CHAPTER 5

1. A study confirming this notion is Seager *et al.* (2006).

NOTES TO CHAPTER 6

1. As argued by Czaja & Frankignoul (2002).

2. Few acronyms are pleasing to the ear, perhaps because they are forced words. Think how ugly the acronyms ENSO, QUANGO, SAT, NASDAQ, GIF, JPEG, and MS-DOS are. A few exceptions, notably *radar* and *laser*, do have a certain euphony and have become accepted as true words with no need for capitalization.

3. The name El Niño was originally used by fishermen along the coasts of Ecuador and Peru to refer to a warm ocean current that usually appears around Christmas and lasts for several months. These days, the name has come to refer only to those events in which the warming is particularly strong.

4. For example, the definition given by Trenberth (1997) is that an El Niño occurs if the five-month running means of sea-surface temperature (SST) anomalies in the Niño 3.4 region (5° N–5° S, 120° W–170° W) exceed 0.4°C for six months or more.

5. Figure kindly provided by A. Wittenberg. The observations are mainly from satellites and are passed through a NOAA optimal interpolation analysis.

6. Figure kindly provided by Andrew Wittenberg, with data from http://www.cgd.ucar.edu/cas/catalog/climind/soi.html. See also Wittenberg (2009). The SOI index is probably quantitatively unreliable before about 1935.

7. See Tudhope et al. (2001) for discussion of a long coral record, and D'Arrigo and Jacoby (1991) for a tree-ring record.

8. Gilbert Walker (1868–1958) may also be credited with discovery of the Southern Oscillation and the North Atlantic Oscillation. Walker began his career as an applied mathematician but made greater contributions, and gained greater fame, by analyzing meteorological observations.

9. The onset of an El Niño involves the westward propagation of Rossby waves and their reflection at the western boundary and return as Kelvin waves. More detailed accounts are to be found in books by Philander (1990), Clarke (2008), and Sarachik and Cane (2010).

NOTES TO CHAPTER 7

1. Brohan et al. (2006) provide a detailed description of the data used to produce the HadCRUT3 data set, one of the most widely used records of temperature of the past 150 years. Another widely used temperature record has been constructed by NASA at their Goddard Institute for Space Studies laboratory, and these data were

used to construct figure 7.1. The stations that NASA uses overlap with those used to construct the HadCRUT record, but the methods differ in how they interpolate the temperature over regions with few records. The resulting two temperature records are similar but not identical.

2. For example, Thompson et al. (2008).

3. Accounting for errors caused by orbital decay has been particularly troublesome, and early analyses of the temperatures that did not properly take it into account did not show a temperature trend that agreed with the surface measurements. When this and other calibration issues are properly accounted for, the temperature trends from satellites agree well with those of the surface measurements.

4. The original "hockey stick" calculation was presented by Mann et al. (1998) and featured prominently in IPCC's third assessment report. For further criticism, discussion, and analysis of the calculations, see, for example, McIntyre & McKitrick (2005), Huybers (2005), Wahl & Ammann (2007), and vonStorch et al. (2009). The IPCC is an intergovernmental review body (see glossary).

5. The measurements of CO_2 by C. D. Keeling and colleagues are described in Keeling et al. (2001).

6. Indermuhle et al. (1999) provide one historical record. Estimates vary slightly, depending on the ice core and time period analyzed.

7. These data are based on University of East Anglia/ U.K. Met Office and NASA/GISS analyses of global temperature. There are slight differences in the temperatures

of individual years in the two analyses, but the decadal figures are almost the same.

8. See Lean & Rind (1998) and Solomon et al. (2007) for more discussion about the solar cycle and its effect on climate.

9. For further discussion of this idea, see Archer (2010).

10. The estimates of the climate sensitivity given in this chapter are based on Meehl et al. (2007) and Padilla et al. (2011).

11. An intercomparison of various climate models is described by Stouffer et al. (2006).

12. The IPCC view, giving a range of sea-level increases from about 20 cm to 60 cm in the coming century, is described in Meehl et al. (2007). The possibility that this is an underestimate is discussed by, for example, Rahmstorf (2010).

13. See also Comiso et al. (2008) and references therein.

Further Reading

MAINLY THE ATMOSPHERE

Wallace, J. M. & Hobbes, P., 2006. *Atmospheric Science: An Introductory Survey.* 2d ed. Burlington, Mass., Academic Press. Covers a wide range of topics at the advanced undergraduate level.

Andrews, D. G., 2010. *An Introduction to Atmospheric Physics.* 2d ed. Cambridge, U.K., Cambridge Univ. Press. Written at the upper-division undergraduate level, with a couple of chapters on dynamics.

Holton, J. R., 2004. *An Introduction to Dynamic Meteorology.* 4th ed. Burlington, Mass., Academic Press. A textbook on dynamics at the undergraduate and graduate levels.

MAINLY THE OCEAN

Denny, M., 2008. *How the Ocean Works: An Introduction to Oceanography.* Princeton, N.J., Princeton Univ. Press. Discusses the ocean from a mechanistic point of view, including physical, chemical, and biological aspects.

An Open University Course Team, 1998. *The Ocean Basins: Their Structure and Evolution.* 2d ed. Oxford, U.K., Pergamon Press.

An Open University Course Team, 2001. *Ocean Circulation.* 2d ed. Oxford, U.K., Pergamon Press. The Open University has a series of books on various aspects of earth sciences written at the undergraduate level.

Pickard, G. L. & Emery, W. J., 1988. *Descriptive Physical Oceanography: An Introduction.* 5th ed. Oxford, U.K., Butterworth-Heinemann. A descriptive survey of the oceans, ocean basins, seawater, and circulation.

Knauss, J. A., 1997. *Introduction to Physical Oceanography.* 2d ed. Long Grove, Ill., Waveland Press. Covers some of the same ground as Pickard and Emery, but with more emphasis on the underlying physical principles and dynamics.

CLIMATE

Bigg, G., 2003. *The Oceans and Climate.* 2d ed. Cambridge, U.K., Cambridge Univ. Press. Goes beyond this book by covering the chemical and biological, as well as physical, interactions.

Kump, L. R., Kasting, J. F. & Crane, R. G., 2009. *The Earth System.* 3d ed. Prentice Hall. A book on the Earth system as a whole, from paleoclimate and ecosystems to ocean circulation and global warming.

Hartmann, D. L., 1994. *Global Physical Climatology*. San Diego, Calif., Academic Press. Covers the physical principles of the climate system, from the fundamental principles to how the global climate system works as a whole.

Marshall, J. C. & Plumb, R. A., 2008. *Atmosphere, Ocean, and Climate Dynamics: An Introductory Text*. Burlington, Mass., Elsevier Academic Press. Covers a variety of topics in climate dynamics at a level appropriate for advanced undergraduates or beginning graduate students.

http://www.realclimate.org/ This website has many informative, and sometimes pointed, blogs on climate issues. As with any website, the links may not last indefinitely.

This glossary gives an informal description of some of the main technical terms used in the book.

Abyss: The deep ocean, extending from the thermocline to the seafloor. Sometimes, the abyss is taken to be only the deepest part of that region, with the middepth ocean extending from immediately beneath the thermocline to about 2 km deep and with the abyss beneath this.

Aerosols: Particulate matter in the atmosphere with both anthropogenic origins (e.g., pollution) and natural origins (e.g., volcanoes).

Baroclinic instability: A hydrodynamic instability that gives rise to weather in the atmosphere and to mesoscale eddies in the ocean.

Carbon dioxide, CO_2: A trace gas, consisting of two oxygen atoms bonded to one carbon atom, comprising about 0.039% of the atmosphere. It is a major greenhouse gas, second only to water vapor in its effect in the current atmosphere, but it is the single most important cause of the greenhouse effect.

Centrifugal force: An apparent force that acts on all bodies in a rotating frame of reference. The force acts to push bodies away from the axis of rotation, but is generally small compared to gravity in Earth's atmosphere and ocean.

Climate sensitivity: A measure of how the climate responds to increasing concentrations of anthropogenic greenhouse gases. Often defined as the average surface temperature response to a doubling in CO_2 from preindustrial levels. The *equilibrium* climate sensitivity is the response when all aspects of the climate system, and in particular the ocean, have equilibrated. The *transient* climate response is the temperature increase that has occurred at the time of CO_2 doubling, usually supposing CO_2 levels to increase by 1% per year. The equilibrium response will be larger than the transient response.

Clouds: Suspended masses of water droplets and/or small ice crystals. Clouds are generally formed when a mass of water vapor condenses, often when moist air is lifted and cools.

Convection: A fluid motion that is caused by buoyancy variations, usually leading to vertical motion—sinking of dense fluid and rising of light fluid.

Conveyor belt: The name given to the global-scale circulating pathway of water in the ocean, generalizing the meridional overturning circulation to multiple basins and including components of the wind-driven circulation. The conveyor belt is a metaphor, albeit a useful one, because the ocean is a turbulent fluid and there is no smooth pathway of fluid from basin to basin.

Coriolis force: An apparent force that acts on bodies that are in motion in a rotating frame of reference. The force acts to the right of the motion in the Northern Hemisphere and to the left in the Southern Hemisphere.

Differential rotation: The effect of Earth's sphericity and rotation combine in such a way that the Coriolis parameter increases from the equator to the pole. This effect is similar to that which arises in differentially rotating systems, in which the actual rotation rate varies with distance from the axis of rotation.

Diffusion: *Molecular diffusion* is the mixing of properties of a fluid by the quasi-random motion of its molecules. *Turbulent diffusion* is mixing by the quasi-random motion of small parcels of fluid and is usually much more efficient than molecular diffusion.

Ekman flow: The flow in the upper ocean caused by the action of the wind and its deflection by the Coriolis force. The net Ekman flow is at right angles to the direction of the wind. The Ekman layer is the layer in the upper ocean, typically a few tens of meters thick, in which the Ekman flow occurs. (There is also a weak Ekman layer at the bottom of the ocean caused by friction.)

El Niño and the Southern Oscillation (ENSO): El Niño is the appearance of anomalously warm water in the eastern equatorial Pacific Ocean, usually lasting for periods of six months to a year. The Southern Oscillation is an associated atmospheric pressure variation, originally defined as the pressure difference between Darwin and Tahiti. El

Niño and the Southern Oscillation vary synchronously, and ENSO refers to the two phenomena together.

Geostrophic balance: A force balance between the Coriolis force and the pressure gradient force that holds to a good approximation on the large scales in the horizontal in both ocean and atmosphere.

Greenhouse effect: The warming effect of greenhouse gases on the surface of Earth. Without this effect, Earth's surface would have a temperature of about 255 K (−18°C), about 33 degrees lower than it actually is.

Greenhouse gas: A greenhouse gas is a gas that absorbs and re-emits infrared or longwave radiation and so keeps Earth's surface warmer than it would be in the gas's absence. The main greenhouse gases are water vapor, carbon dioxide, ozone, nitrous oxide, and methane (see table 1.2). Water vapor differs from the other greenhouse gases in that its overall level in the atmosphere is greatly influenced by temperature and it is regarded as a feedback rather than a primary "forcing" of the greenhouse effect.

Gyres: Gyres are the great circulating water masses in the ocean basins, stretching across the ocean basins from one continent to another. They are largely driven by the overlying winds.

Halocline: See Thermocline.

Holocene: The period since the end of the last ice age, and so about the past 12,000 years.

Hydrostatic balance: A force balance between gravity and the pressure gradient force that normally holds to a good approximation in the vertical direction, in both ocean and atmosphere.

Infrared radiation or longwave radiation: Electromagnetic radiation with wavelengths rather larger than those of visible light. Such radiation is emitted by Earth's surface and atmosphere (whereas the sun emits mostly visible radiation).

Intergovernmental Panel on Climate Change (IPCC): Intergovernmental body established in 1988 by the World Meteorological Organization and the United Nations Environment Programme and charged with reviewing and assessing the scientific and socioeconomic knowledge relevant to climate change. It has produced assessment reports in 1990, 1995, 2001, and 2007, and a fifth report is due in 2014.

Isopycnal: Of constant density. An isopycnal surface is a surface on which density is constant, although salinity and temperature might vary.

Meridional overturning circulation: The large-scale circulation of the ocean that occurs in the meridional (north–south, up–down) plane. The circulation has a pole-to-pole interhemispheric component, mainly in the Atlantic Ocean, as well as pole-to-equator components in the major basins.

Mesoscale eddy: A type of variability, or eddy, in the ocean with a typical horizontal scale between about

50 km and 300 km and a timescale of weeks. These eddies produce variability that is analogous to that of weather in the atmosphere. The majority of the kinetic energy of the ocean resides in such eddies rather than in the more sluggish mean circulation.

Mixed layer: The uppermost layer of the ocean, in which properties such as temperature and salinity are well mixed and so do not change in the vertical. (The values may vary horizontally.) Typically the mixed layer is 50–100 m thick, although it can be much thicker in convective regions.

North Atlantic Oscillation: The apparent north–south oscillation of the weather pattern in the North Atlantic, particularly in winter. The pattern oscillates on timescales of days to weeks, but some years are characterized by the preponderance of the pattern in one phase much more than the other.

Positive feedback: A process in which a small change is amplified by some effect, leading to a larger change, which in turn is amplified more, and so on. To be contrasted with a negative feedback, in which the initial change is damped.

Pycnocline: See Thermocline.

Salinity: Salinity is the dissolved salt or mineral concentration of water. In the ocean the salinity is mainly a combination of chloride (55% of the total salt), sodium (31%), sulfate (7.7%), magnesium (3.7%), calcium (1.2%), and potassium (1.1%), in virtually the same ratio to each other throughout the ocean. Over most of the ocean, the salinity lies between 32 and 37 parts per thousand (g/kg). It is often determined by measuring the electrical conductivity of a water sample.

Sea Ice: Ice that is formed by the freezing of seawater.

Southern Oscillation: See El Niño and the Southern oscillation.

Sverdrup: A unit of transport. Originally defined to be a volume transport, 10^6 m^3 s^{-1} of seawater, but perhaps more usefully thought of as a mass transport, 10^9 kg s^{-1}, for it may then be used in the atmosphere as well as the ocean.

Sverdrup balance: The balance between the curl of the forces produced by wind stress and by a meridional flow in the presence of differential rotation. Mathematically represented by curl $\tau = \beta v$, where τ is the wind stress and $\beta = \partial f / \partial y$.

Thermocline: The region of the upper ocean in which temperature changes rapidly in the vertical, connecting the mixed layer with the abyss. The thermocline is typically 500–1,000 m thick. The *pycnocline* and the *halocline* are regions where the density and salinity change rapidly and often coincide with the thermocline.

Thermohaline circulation: A name that is sometimes given to the large-scale overturning circulation in the ocean. The name suggests that temperature and salinity are the main influences on the circulation, but they are not the only influences; the more generic appellation *meridional overturning circulation* is sometimes preferred, unless the thermal and haline effects are being specifically referred to.

Water vapor: The gaseous state of water, not to be confused with steam, which (like clouds) is composed of water droplets. Water vapor is the main greenhouse gas, but its overall level is determined by temperature and so it is not a primary forcing of the greenhouse effect.

Western boundary current: The main oceanic gyres' intense boundary currents in the west, such as the Gulf Stream in the Atlantic and the Kuroshio in the Pacific, produced by the differential rotation of Earth (see Differential rotation).

Western intensification: The intensification of the currents in the western parts of the gyres, leading to the production of western boundary currents.

References

Archer, D., 2010. *The Long Thaw*. Princeton Univ. Press, 192 pp.

Bindoff, N., Willebrand, J., Artale, V., Cazenave, A. et al., 2007. Observations: Ocean climate change and sea level. In S. Solomon, D. Qin, M. Manning, Z. Chen, M. Marquis, K. Averyt, M. Tignor & H. Miller, Eds., *Climate Change 2007: The Physical Science Basis. Contribution of Working Group I to the Fourth Assessment Report of the Intergovernmental Panel on Climate Change*, pp. 335–432. Cambridge, U.K., and New York: Cambridge University Press.

Brohan, P., Kennedy, J. J., Harris, I., Tett, S.F.B. & Jones, P. D., 2006. Uncertainty estimates in regional and global observed temperature changes: A new dataset from 1850. *J. Geophys. Res.*, 111, D12106, doi:10.1029/2005JD00654.

Chelton, D. B., Schlax, M. G. & Samelson, R. M., 2011. Global observations of nonlinear mesoscale eddies. *Prog. Oceanography*, in press.

Clarke, A. J., 2008. *Dynamics of El Niño and the Southern Oscillation*. Academic Press, 308 pp.

Comiso, J., 2002. A rapidly declining perennial sea ice cover in the Arctic. *Geophys. Res. Lett.*, **29**.

Comiso, J. et al., 2008. Accelerated decline in the Arctic sea ice cover. *Geophys. Res. Lett.*, **35**, L01703.

Czaja, A. & Frankignoul, C., 2002. Observed impact of Atlantic SST anomalies on the North Atlantic Oscillation. *J. Climate*, **15**, 606–23.

D'Arrigo, R. & Jacoby, G. C., 1991. A 1000-year record of winter precipitation from northwestern New Mexico, USA: A reconstruction from tree-rings and its relation to El Niño and the Southern Oscillation. *The Holocene*, **1**, 95–101.

Domingues, C. M., White, J. A., Church, N. J., Gleckler, P. J., Wijffels, S., E., Barker, P. M., & Dunn, J. R., 2008. Improved estimates of upper-ocean warming and multi-decadal sea-level rise. *Nature*, **453**, 1090–94.

Houghton, J., Ding, Y., Griggs, D., Noguer, M. et al., Eds., 2001. *IPCC, 2001: Climate Change 2001: The Scientific Basis. Contribution of Working Group I to the Third Assessment Report of the Intergovernmental Panel on Climate Change.* Cambridge University Press, 996 pp.

Huybers, P., 2005. Comment on "Hockey sticks, principal components, and spurious significance" by S. McIntyre & R. McKitrick. *Geophys. Res. Lett.*, **32**, L20705.

Indermühle, A., Stocker, T. F., Joos, F., Fischer, H. et al., 1999. Holocene carbon-cycle dynamics based on CO_2

trapped in ice at Taylor Dome, Antarctica. *Nature*, **398**, 121–26.

Jones, P. D. & Mann, M. E., 2004. Climate over past millennia. *Rev. Geophys.*, **42**, RG2002, doi:10.1029/2003RG000143.

Keeling, C. D., Piper, S. C., Bacastow, R. B., Wahlen, M. et al., 2001. Exchanges of atmospheric CO_2 and $^{13}CO_2$ with the terrestrial biosphere and oceans from 1978 to 2000. I. Global aspects. Technical report, SIO reference series, Scripps Institution of Oceanography, San Diego.

Kielh, J. T. and Trenberth, K. E., 1997. Earth's annual global mean energy budget. *Bull. Am. Meteor. Soc.*, **78**, 197–208.

Lacis, A., Schmidt, G. A., Rind, D. & Ruedy, R. A., 2010. Atmospheric CO_2: Principal control knob governing Earth's temperature. *Science*, **330**, 356–59.

Lean, J. & Rind, D., 1998. Climate forcing by changing solar radiation. *JC*, **11**, 3069–92.

Lozier, M. S., Owens, W. B. & Curry, R. G., 1995. The climatology of the North Atlantic. *Prog. Oceanography*, **36**, 1–44.

Mann, M. E., Bradley, R. S. & Hughes, M. K., 1998. Global-scale temperature patterns and climate forcing over the past centuries. *Nature*, **392**, 779–87.

McIntyre, S. & McKitrick, R., 2005. Hockey sticks, principal components, and spurious significance. *Geophys. Res. Lett.*, **32**, L03710.

Meehl, G., Stocker, T., Collins, W., Friedlingstein, P. et al., 2007. Observations: Global climate projections. In S. Solomon, D. Qin, M. Manning, Z. Chen, M. Marquis, K. Averyt, M. Tignor & H. Miller, Eds., *IPCC, 2007: Climate Change 2007: The Physical Science Basis. Contribution of Working Group I to the Fourth Assessment Report of the Intergovernmental Panel on Climate Change*, pp. 747–846. Cambridge and New York: Cambridge University Press.

Myhre, G., Highwood, E., Shine, K. & Stordal, F., 1998. New estimates of radiative forcing due to well mixed greenhouse gases. *Geophys. Res. Lett.*, **25**, 14, 2715–18.

Padilla, L., Vallis, G. K. & Rowley, C., 2011. Probabilistic estimates of transient climate sensitivity subject to uncertainty in forcing and natural variability. *J. Climate.* In press.

Pedlosky, J., 1987. *Geophysical Fluid Dynamics*, 2nd ed. Springer-Verlag, 710 pp.

Persson, A., 1998. How do we understand the Coriolis force? *Bull. Am. Meteor. Soc.*, **79**, 1373–85.

Philander, S. G., 1990. *El Niño, La Niña, and the Southern Oscillation*. Academic Press, 289 pp.

Rahmstorf, S., 2010. A new view on sea level rise. *Nature Reports: Climate Change*, **4**, 44–45.

Sarachik, E. S. & Cane, M. A., 2010. *The El Niño–Southern Oscillation Phenomenon*. Cambridge University Press, 369 pp.

Schmidt, G. A., Ruedy, R. A., Miller, R. L. & Lacis, A. A., 2010. Attribution of the present-day total greenhouse effect. *J. Geophys. Res.*, **115**, D20106.

Seager, R., Battisti, D. S., Yin, J., Gordon, N. et al., 2006. Is the Gulf Stream responsible for Europe's mild winters? *Quart. J. Roy. Meteor. Soc.*, **128**, 2563–86.

Solomon, S., Qin, D., Manning, M., Chen, Z. et al., Eds., 2007. *IPCC, 2007: Climate Change 2007: The Physical Science Basis. Contribution of Working Group I to the Fourth Assessment Report of the Intergovernmental Panel on Climate Change.* Cambridge, U.K., and New York, Cambridge University Press, 996 pp.

Stouffer, R. J., Yin, J., Gregory, J. M., Dixon, K. W. et al., 2006. Investigating the causes of the response of the thermohaline circulation to past and future climate changes. *J. Climate*, **19**, 1365–87.

Thompson, D.W.J., Kennedy, J. J., Wallace, J. M. &Jones, P. D., 2008. A large discontinuity in the mid-twentieth century in observed global-mean surface temperature. *Nature*, **453**, 646–49.

Trenberth, K. E., 1983. What are the seasons? *Bull. Am. Meteor. Soc.*, **64**, 1276–82.

Trenberth, K. E., 1997. The definition of El Niño. *Bull. Am. Meteor. Soc.*, **78**, 2771–77.

Trenberth, K. & Caron, J., 2001. Estimates of meridional atmosphere and ocean heat transports. *J. Climate*, **14**, 2771–77.

Tudhope, A. W., Chilcott, C. P., McCulloch, M. T., Cook, E. R. et al., 2001. Variability in the El Niño Southern Oscillation through a glacial–interglacial cycle. *Science*, **291**, 1511–17.

Vallis, G. K., 2006. *Atmospheric and Oceanic Fluid Dynamics*. Cambridge University Press, 745 pp.

von Storch, H., Zorita, E. & González-Rouco, F., 2009. Assessment of three temperature reconstruction methods in the virtual reality of a climate simulation. *Int. J. Earth Sci.* (*Geol Rundsch*), **98**, 67–82.

Wahl, E. R. & Ammann, C. M., 2007. Robustness of the Mann, Bradley, Hughes reconstruction of Northern Hemisphere surface temperatures: Examination of criticisms based on the nature and processing of proxy climate evidence. *Climatic Change*, **85**, 33–69.

Winton, M., 2003. On the climatic impact of ocean circulation. *J. Climate*, **16**, 2875–89.

Wittenberg, A. T., 2009. Are historical records sufficient to constrain ENSO simulations? *Geophys. Res. Lett.*, **36**, L12702, doi:10.1029/2009GL038710.

World Ocean Atlas 2009, 2009. National Oceanographic Data Center, Silver Spring, Md. <http://www.nodc.noaa.gov/OC5/WOA09/pr_woa09.html>.

Index